SOLE DISTRIBUTORS FOR THE UNITED STATES AND CANADA
AMERICAN ELSEVIER PUBLISHING COMPANY, INC.
52 Vanderbilt Avenue, New York 17, N.Y.

TOPICS IN CHEMICAL PHYSICS

BASED ON THE HARVARD LECTURES OF

PETER J. W. DEBYE

by

ALFRED PROCK

Assistant Professor of Chemistry,
Boston University, Boston, Mass.

and

GLADYS McCONKEY

Technical Assistant,
Cornell University, Ithaca, N.Y.

ELSEVIER PUBLISHING COMPANY

AMSTERDAM - NEW YORK

1962

Library of Congress Catalog Card Number 61–18140

With 60 illustrations

PRINTED IN THE NETHERLANDS BY
N.V. DRUKKERIJ D. REIDEL - DORDRECHT

Preface

A few years ago, at the invitation of Harvard University's chemistry department, Professor P. Debye gave a special series of lectures in chemical physics for graduate students. For these lectures he organized a course in special topics which, while of fundamental importance, are not usually fully covered in graduate programs of study. The subject matter he discussed is highly diversified, but all of it is concerned with areas of chemical physics in which Professor Debye has made significant contributions over a period of many years.

Because of the uniqueness of this assembly of topics, it seemed desirable to have it made available in printed form. The authors were asked by Professor Debye to present the material in a form suitable for publication as a book. This book, then, is a result of Professor Debye's Harvard lectures.

The opening chapter begins with a discussion of the fundamental laws of the static electric field, and how they are generalized to include the effect of the presence of dielectric material. A development of these ideas leads to derivation of expressions for thermodynamic properties of material in the presence of an electrostatic field. The concept of permanent and induced dipole moment is introduced, and from a classical point of view the behavior of a dipolar gas in the presence of an electrostatic field is quantitatively described.

Chapter 2 is concerned with the fundamental concepts of statistical mechanics. Aspects of the convenient but limited classical approach and the rigorous but complicated modern approach are discussed.

The statistical method is put to use in Chapter 3, in which dielectric properties of condensed phases are considered. The chapter begins with a discussion of what information can be obtained about molecular interaction in condensed media from measurements of static dielectric constant.

This leads to a discussion of the reaction of polar molecules in a viscous medium to an alternating field, and then with the aid of quantum mechanics the reaction of a polar gas to both a static and an alternating field is explained.

Chapter 4 presents a discussion of the application of the technique of X-ray and light scattering to the study of particle size and shape and also of over-all structural characteristics of material as expressed by correlation function. Materials considered are crystals, colloids, polymers, and porous solids.

Chapter 5 is concerned with the theory of dilute solutions of strong electrolytes, the well-known Debye-Hückel theory. The thermodynamic properties of such systems as predicted by this theory are compared with those predicted by dissociation theory. The concept of relaxation of the ionic atmosphere is introduced and used to explain certain high frequency and high field strength phenomena observed in the measurement of conductivity of such systems.

The final chapter deals with high molecular weight molecules which can be described as random coils. The discussion includes the calculation of coil size through the interpretation of data obtained by means of the light scattering technique.

Boston, Mass. / Ithaca, N.Y. A. P
April, 1962 G. McC.

Contents

CHAPTER 1

Fundamentals of the Static Electric Field; Dielectric Constant and Polarizability of Gases

The reaction of a material to the influence of an electric field is a measurable property which can reveal a great deal about the molecular structure of the substance. For example, measurements of dielectric constant, a quantity that conveniently characterizes this reaction, can be used to distinguish certain stereoisomers. But in order to make use of measurements of field-induced phenomena, we must understand the fundamental laws of the static field. A static electric field in a vacuum is described by certain well known equations; we shall examine these and see how they must be modified when there is material present.

In order to discuss the subject quantitatively, a system of units is needed. We know from Coulomb's law that the force between two point charges in a vacuum is proportional to the product of the charges divided by the square of the distance between them. If the force is expressed in dynes and the distance in centimeters, and if the proportionality constant is chosen to equal one, we can use Coulomb's law to define the unit of charge such that two unit charges separated by a distance of one centimeter exert on each other a mutual repulsion of one dyne. This charge unit is called the electrostatic unit (esu) of charge, and other quantities derived from it are also expressed in electrostatic units. Coulomb's law, then, is written $F = q_1 q_2/r^2$, with F in dynes, r in centimeters, and the charges, q, in esu. In discussing the mutual effects of charges, it is convenient to use the concept of the electrostatic field: each charge produces a kind of disturbance in the space around it which is experienced as a force by any other charge placed in the vicinity. The electrostatic field strength is defined at every point as the force acting on a unit charge placed at that point. For the case of a point charge in a vacuum, the field E at any place is equal to q/r^2, where r is the distance from the charge q.

References p. 39

The Static Electric Field in Vacuum

Consider now a field set up by a collection of static charges in a vacuum (*i.e.*, with no material bodies present). The results of two simple experiments provide the basis for a complete mathematical description of the electrostatics of the system.

First, it has always been found that the total work done in moving a charge around any closed path in a field is zero. This simple statement has a powerful implication: it implies that the amount of work done in moving a charge from one point to another in a field depends only on the end points and not at all on the path. One can see how this follows by considering any two paths from A to B. The experiment shows that the work required to go from A to B on path one, plus the work from B to A on path two, equals zero. This means that the work on path two, from B to A, must be equal and opposite to that on path one, from A to B. If now the direction is reversed, so that the charge is moved from A to B on path two, we see that the work done is exactly the same as it was on path one, since reversing the direction of travel reverses the sign of work.

In the second experiment, a measurement is made of the electric field produced at the boundary of a given volume by the contained charge. The measured quantity is the integrated value of the normal component of field strength over the enclosing surface. It has always been found that this value is equal to $4\pi Q$, where Q is the total enclosed charge.

Now let us develop the results of these experiments. From the first experiment we know that the net work required to move a charge around a closed path is zero. Expressed mathematically this is:

$$\oint E_s \, \mathrm{d}s = 0 \, ,$$

where $\mathrm{d}s$ is an element of path length, E_s is the projection of the field onto $\mathrm{d}s$, and the integration is performed around the closed path. If the enclosed area and the elements of length are small, the equation can also be written in terms of derivatives of the field (the method is discussed in the Appendix). In the differential expression, we let X, Y, and Z be the three components of the field in a Cartesian system of coordinates x, y, and z; then:

$$\frac{\partial Z}{\partial y} - \frac{\partial Y}{\partial z} = 0 \, , \; \frac{\partial Z}{\partial x} - \frac{\partial X}{\partial z} = 0 \, , \; \frac{\partial Y}{\partial x} - \frac{\partial X}{\partial y} = 0 \, ;$$

or, in more abbreviated form:

$$\text{curl } \tilde{E} = 0\,.$$

We also saw from the first experiment that the work done in moving a unit charge in an electrostatic field is independent of the path and depends only on the end points. Accordingly, there must be a function Φ, dependent on position only, such that the work done from point A to point B is given by $\Phi(\mathrm{B}) - \Phi(\mathrm{A})$. Because we are not concerned with energy loss due to friction, and since the charge is assumed to move slowly enough so that kinetic energy is negligible, this work appears as a change in potential energy and Φ is therefore called the potential. Its three partial derivatives give (except for sign) the three components of the force on the charge:

$$X = -\frac{\partial \Phi}{\partial x}\,,\ Y = -\frac{\partial \Phi}{\partial y}\,,\ Z = -\frac{\partial \Phi}{\partial z}\,.$$

The minus signs are included in order to make the expressions conform to the idea that one must push the charge against the force due to the field in order to raise its potential energy. The potential Φ is a solution to the set of differential equations summarized by the expression curl $\tilde{E} = 0$, as can be verified by substitution. These points are discussed in greater detail in the Appendix.

The second experimental law is expressed in the form of an equation as:

$$\int_s E_n \, \mathrm{d}\sigma = 4\pi Q\,,$$

where E_n is the normal component of field strength and $\mathrm{d}\sigma$ is the surface element. This is known as Gauss's law. In the simple case of an isolated point charge, the equivalence of Gauss's law and Coulomb's law is readily apparent: we know from Coulomb's law that E_n for the point charge is q/r^2, and so the integral, taken over a sphere of any radius, is evaluated as $4\pi q$. In the general equation, the appearance of total charge Q implies that total field at any point is just the superposition of the fields due to individual charges.

A pictorial representation of the experimental law can be made in terms of the concept of lines of force. We picture the field around a charge distribution as a number of lines of force emanating from the enclosing volume. A line of force can be described in the following way. When a

References p. 39

positive charge is put into an electrostatic field at a point A, it will experience a force. If the charge is moved a small step in the direction of the force to a new point it will, in general, experience a new force. If this process is continued to point B and the limiting case in which the steps are very small is taken, we can say that a line of force from point A to point B has been described. Field direction at any point is given by the tangent to the line of force at the point. It can be seen that the field continuously does work on the charge in moving it along a line of force, and since according to our first experimental result the net work around a closed path is zero, we must conclude that a line of force cannot close upon itself.

The strength of the field can be indicated in the pictorial representation by making the density of lines of force at some point correspond numerically to the field strength there. This means that the number of lines passing through a one square centimeter area is numerically equal to the field strength normal to that area. Thus the total number of lines of force emanating from a given volume is given by the integral $\int_s E_n \, d\sigma$; we know from the second experimental law that this is equal to $4\pi Q$.

Gauss's law may also be written in a differential form, using the Cartesian coordinate system:

$$\frac{\partial X}{\partial x} + \frac{\partial Y}{\partial y} + \frac{\partial Z}{\partial z} = 4\pi\rho \, ,$$

where ρ is charge per cubic centimeter. In more abbreviated form:

$$\text{div}\, \tilde{E} = 4\pi\rho \, .$$

Details of the development of this equation are given in the Appendix. A general interpretation in terms of our pictorial representation of field as lines of force, is that in a finite region a line of force can begin or end only on an element of charge.

If we rewrite this equation, substituting for div $\tilde{E}$ the expressions for field components we derived from the first experiment, we obtain the general differential equation for electrostatics:

$$\triangle \, \Phi \equiv \frac{\partial^2 \Phi}{\partial x^2} + \frac{\partial^2 \Phi}{\partial y^2} + \frac{\partial^2 \Phi}{\partial z^2} = - \, 4\pi\rho \, ,$$

which is called Poisson's equation. In the special case when $\rho = 0$ (*i.e.*, when the region under consideration is free of charge):

$$\triangle \Phi = 0 .$$

This is called Laplace's equation.

Poisson's equation (or in the case of a charge-free region, Laplace's equation) with the appropriate boundary conditions, constitutes the complete statement of the electrostatic problem when no material bodies are present. Once the potential Φ has been expressed in the coordinates chosen, simple differentiation yields the components of the field at any point.

The Static Electric Field in the Presence of Material

Now let us turn our attention to the results of a very important experiment performed by Faraday. He found that the charge acquired by a condenser (two parallel metal plates separated by an insulator) at a given potential difference, depended on the material between the plates; in particular, he found that the smallest charge was acquired when vacuum separated the plates. For a vacuum condenser the relation was found to be $Q = C\Phi_0$, where Q represents the charge on one of the plates, Φ_0 is the difference of potential between them, and C is a proportionality constant depending on the geometry of the system*. When material was placed between the plates, a different charge, Q', was acquired at the same potential difference.

The ratio Q'/Q, which is always greater than unity, is termed the dielectric constant, ε. Dielectric constant may be combined with the proportionality constant C to define a new constant, C', called capacity: $C' = \varepsilon C$. Thus the expression for charge becomes $Q' = C'\Phi_0$.

The dielectric constant of a vacuum is unity, and for most gases it is just a few tenths of a percent higher at ordinary conditions of temperature and pressure. But dielectric constant can be quite high. For many solids the order of magnitude is around 5; for liquid water ε is 80, and for sulfuric acid it has been estimated to be about 10,000.

It was apparent from Faraday's experiments that a field produces some kind of effect on material. Many investigators (including Mossotti, who

* The constant C has the dimension of length and is equal to $S/4\pi d$ where S is the area of one of the plates and d is the distance separating them. A fuller discussion of this constant is given in the Appendix p. 30.

worked out a quantitative theoretical description of the phenomenon) postulated that the external field, though unable to ionize the insulating material, causes a displacement of charges within the molecules. Such a displacement would superimpose an internal field whose effect would be to decrease the macroscopic field if the plate charge remained the same. But the macroscopic field does not decrease since the potential difference between the plates is experimentally maintained after introduction of material. Therefore the effect of charge displacement on field intensity must be offset by additional charge on the plates.

In formulating a quantitative theory describing this phenomenon, Mossotti assumed that a matrix of small conducting spheres with a mutual separation large compared to their radii can be used to describe the material between the condenser plates*. The problem is then one of formal electrostatics: the charge displacement of a tiny sphere must be calculated. It can be shown ** that a sphere of radius a behaves electrically just like a dipole, which is an electrical entity consisting of two equal and opposite charges separated by some distance. Dipole moment, m, is defined as qr, the product of the magnitude of charge and the separation distance; the sphere of radius a corresponds to a dipole with a moment equal to a^3E.

It is convenient to define the proportionality constant between induced dipole moment and field strength as polarizability per molecule, α. Thus:

$$m = \alpha E\,.$$

For our polarizable sphere, α equals a^3, a quantity proportional to the volume of the sphere †.

In experiments with dielectric materials, it is generally the dielectric constant that is directly measurable. Therefore, if we want to find out what the polarizability of a material is we must find some way of relating α to the measured ε. To do this we must first reconsider the experimental laws to discover what modifications are necessary when material is present.

---

* This model is a good one only for gases because interaction between spheres is neglected.

** This is discussed in the Appendix, which also gives preliminary discussions of potential due to a point charge, and potential due to a dipole (pp. 30–33).

† The similarity between a classical model of the atom and a conducting sphere is pointed out in the Appendix (pp. 33–34).

How does the presence of material affect our first law, which concerns the work done in moving a charge around a closed path? Consider such a closed path intersecting the material, and let E be the field strength at any point along it. If all the substance in a vanishingly small needlelike cavity around the path were removed, this would not change the field strength along the path, yet the path would be left entirely in vacuum. In other words, if the field in the needlelike cavity is designated as E_{I}, then $E_{\mathrm{I}} = E$. The net work around the path must accordingly be zero; hence the first experimental law is valid when material is present as well as in vacuum.

A similar technique can be used to determine the number of lines of force passing through a closed surface buried in the medium. Suppose the substance immediately adjacent to both sides of the surface were removed, leaving a thin shell of vacuum containing the surface. The intensity of the field intersecting the surface, designated E_{II}, represents the field strength we require for integration over the enclosing surface. It is related directly to the enclosed charge and, as we shall see, is not generally the same as E_{I} except in vacuum.

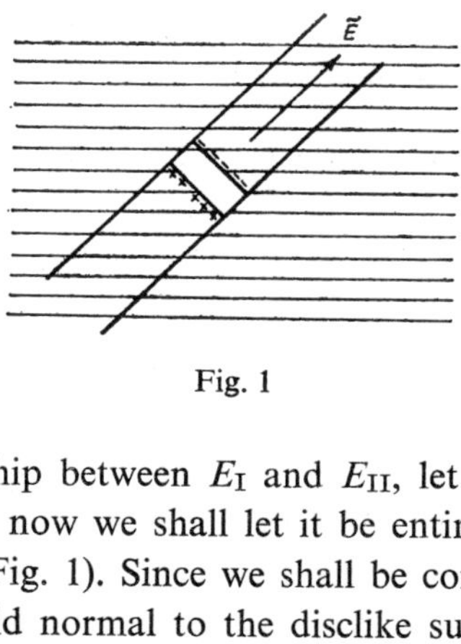

Fig. 1

To find the relationship between E_{I} and E_{II}, let us consider again the needlelike cavity, but now we shall let it be entirely filled except for a disclike volume (see Fig. 1). Since we shall be concerned only with the component of the field normal to the disclike surface, without loss of generality we can assume that the field is perpendicular to the surface. Suppose the surrounding material is polarized and P is the electric dipole moment per unit volume, assumed constant over the cavity. There will be charge residing on the flat surfaces of the disclike volume, for at the surfaces there will be dipole ends whose charges are not counteracted by adjacent oppositely-charged dipole ends. Therefore the field E_{II} in the

References p. 39

disclike volume will be made up of two quantities: the original field $E_{\rm I}$, and the field due to this surface charge.

The surface charge per unit area, σ, is the same numerically and dimensionally as the quantity P, as can be seen through the following consideration. If a unit volume is considered to contain n dipoles, each of unit length, the number of dipole ends per unit area at a surface perpendicular to the direction of dipole orientation must equal the number of dipoles per unit volume, n. If q is the charge of the dipole, then charge per unit area, σ, equals qn. Dipole moment per molecule, m, equals $q \times 1$, and if polarization, P, is assumed to equal mn, then P equals qn. Accordingly, σ and P are equivalent.

The field due to this surface charge can be calculated by means of our second experimental law, $\int_s E_n \mathrm{d}\sigma = 4\pi Q$, for the volume considered is a vacuum. In the disclike cavity the field direction is normal to the surface, and so the field equals $4\pi Q/s$ or $4\pi\sigma$, which is $4\pi P$.

Total field $E_{\rm II}$ is thus $E_{\rm I} + 4\pi P$, or, since $E_{\rm I}$ was shown to be the macroscopic field E, $E_{\rm II}$ equals $E + 4\pi P$. The field $E_{\rm II}$, which is the field strength needed for the integration over the surface enclosing the polarized medium, is designated by a new symbol, D, called the electric displacement. In the presence of material, then, the expression for our second experimental law must be modified to read:

$$\int_s D_n \mathrm{d}\sigma = 4\pi Q, \text{ where } D = E + 4\pi P\,.$$

In summary, the two basic equations applicable to systems in which material is present are:

A. $\oint E_s \mathrm{d}s = 0$ or curl $\tilde{E} = 0$. (This is the same as for the vacuum system.)

B. $\int_s D_n \mathrm{d}\sigma = 4\pi Q$ or div $\tilde{D} = 4\pi\rho$, where $D = E + 4\pi P$†.

These formulas hold not only for polarizable spheres but also for polarizable molecules so long as the material is isotropic. However, molecules lack spherical symmetry, and so the induced moment per molecule, m, is not generally parallel to the field but lies along some preferred molecular axis. This effect would be important in a crystalline solid, but in a gas or liquid where the molecules are free to assume any position, P will be an

† Another method of deriving electric displacement and the differential form of the equation is given in the Appendix (pp. 34–36).

average quantity always in the direction of E. We shall consider only such substances.

We have defined dielectric constant of a polarizable material as the ratio of charge on condenser plates with and without the material between them. But we can express this quantity ε in a more useful form in terms of the applied field and the polarizability of the dielectric material. Our second experimental law, when applied to the conditions of a parallel plate condenser where the field direction is normal to the plates, shows that surface charge density, σ, equals $E/4\pi$ for vacuum and $D/4\pi$ when material is present. Therefore the ratio of charge, which is dielectric constant, is:

$$\varepsilon = D/E\,.$$

Since $D = E + 4\pi P$:

$$\varepsilon = 1 + 4\pi\frac{P}{E}\,.$$

Over the range of field strength ordinarily used in the laboratory, ε is constant for all practical purposes; hence P/E must be constant. The proportionality constant is called the susceptibility, or polarizability per unit volume, and is given the symbol κ, so that:

$$P = \kappa E\,.$$

If we assume that the polarizability per unit volume, κ, equals $n\alpha$, where n is the number of molecules per cubic centimeter, then we can relate the measurable quantity ε to the polarizability of a particle, α:

$$\varepsilon - 1 = 4\pi n\alpha\,.$$

For the case of a conducting sphere, we found that α equals a^3, where a is the radius of the sphere. A very interesting relationship is found by substituting this value for α and dividing both sides of the equation by 3:

$$\frac{\varepsilon - 1}{3} = \frac{4}{3}\pi a^3 n\,.$$

According to this, the quantity $(\varepsilon - 1)/3$ is a measure of the volume occupied by the molecules in a cubic centimeter of material. There is another way to determine the volume occupied by molecules in a volume of gas; based on van der Waals' equation, it depends only on measurements

References p. 39

of pressure, volume, and temperature. When volumes calculated by the two methods are compared, fair agreement is found in some instances, while in others, the volumes calculated through the dielectric constant are much larger. On closer examination one finds that into the second group fall those molecules that have separated centers of positive and negative charge, producing permanent dipole moments. Dielectric constants of systems containing molecules with permanent dipole moments will be discussed later.

Energy Relations in an Electric Field

In order to discuss and predict the gross properties of material in the presence of an electric field we shall make use of the laws of thermodynamics. Concepts of energy and entropy per unit mass or volume are of great utility in formulating thermodynamic relationships, and so we want to develop an expression for the energy contributed to any given unit of volume by any distribution of charge. In other words, we want to calculate total energy density at any point. This should be expressible in terms of the quantities field strength, E, and electric displacement, D, which is εE.

This seems reasonable when we think of a system under the influence of a field in terms of a parallel plate condenser. During the process of charging a condenser, work is done in moving charge to create the potential difference developed between the plates. Accordingly, the stored energy at any time is a function of two quantities, charge and potential difference. But the field strength between the plates is also a function of potential difference, and displacement is a function of charge, so it should be possible to express stored energy as a function of the quantities E and D.

To develop an expression for energy density, let us begin with the assumption of a charge distribution, expressed as charge per unit volume, ρ, existing in space. There is a potential, Φ, existing at every point. We ask how much work is required to change the charge density by a very slight amount, from ρ to $(\rho + \mathrm{d}\rho)$. According to the Coulomb point of view, the work required per unit volume is the product of the potential Φ and the change in charge density $\mathrm{d}\rho$; to obtain total change of energy, one must sum over the volume containing the charge: $\mathrm{d}W = \int_V \Phi \,\mathrm{d}\rho \,\mathrm{d}v$. This equation can be transformed into one involving only the two field quantities E and D (details of the vector transformation are given in the Appendix). The resultant equation is:

$$\mathrm{d}W = \frac{1}{4\pi} \int_{\text{all space}} \tilde{E} \cdot \mathrm{d}\tilde{D} \, \mathrm{d}v \,,$$

which is an expression of the Faraday point of view. We shall make the assumption that the integrand (with the constant $1/4\pi$) represents the change in energy per unit volume, $\mathrm{d}w$. (This is an assumption because $\mathrm{d}W$ is given by a definite integral.) Thus $\mathrm{d}w$ is $\tilde{E} \cdot \mathrm{d}\tilde{D}/4\pi$, or, when $\tilde{E}$ and $\tilde{D}$ lie in the same direction:

$$\mathrm{d}w = E\mathrm{d}D/4\pi \,.$$

For a vacuum, $\mathrm{d}D = \mathrm{d}E$, and simple integration shows that the energy density is given by $E^2/8\pi$; for the special case when ε does not change with temperature, the energy density in material is given by $\varepsilon E^2/8\pi$, where εE is D.

However, for the general consideration of the energy of a system containing material, one must take into account not only electrical energy $\varepsilon E^2/8\pi$, but also a 'hidden energy', which is the result of the thermal variation of dielectric constant. For the general case, then, the change in energy density is:

$$\mathrm{d}w = \frac{E\mathrm{d}D}{4\pi} = \frac{\varepsilon E\mathrm{d}E}{4\pi} + \frac{E^2\mathrm{d}\varepsilon}{4\pi} \,.$$

If we designate the symbol ω as the vacuum energy density $E^2/8\pi$, then $\mathrm{d}\omega = E\mathrm{d}E/4\pi$, and upon substituting this into our equation we obtain:

$$\mathrm{d}w = \varepsilon\mathrm{d}\omega + 2\omega\mathrm{d}\varepsilon \,.$$

This expression for energy density can be incorporated into thermodynamic equations as the electrical contribution to the work term. The combined first and second laws of thermodynamics state that at equilibrium $\mathrm{d}U = \mathrm{d}W + T\mathrm{d}S$, where $\mathrm{d}U$ is the change in internal energy, $\mathrm{d}W$ is the work done on the system, and $T\mathrm{d}S$ represents the heat put into the system, $\mathrm{d}S$ being the change in entropy. If the volume is kept constant, the nature of the work is electrical only, and so on a unit volume basis $\mathrm{d}U = T\mathrm{d}S + \varepsilon\mathrm{d}\omega + 2\omega\mathrm{d}\varepsilon$. Under the influence of electric fields in the

References p. 39

range usually used in the laboratory, dielectric constant is a function of temperature only (at constant volume), and so the quantity dε may be expressed as $(\mathrm{d}\varepsilon/\mathrm{d}T)\mathrm{d}T$. Subtracting the total differential $\mathrm{d}(TS)$ from both sides of the equation, we obtain:

$$\mathrm{d}(U - TS) = \left(-S + 2\omega \frac{\mathrm{d}\varepsilon}{\mathrm{d}T}\right) \mathrm{d}T + \varepsilon \mathrm{d}\omega \,.$$

The function on the left hand side is sometimes defined as dF, where F is a state function called the Helmholtz free energy or thermodynamic potential. Since dF is a total differential, it is possible to relate the differentials on the right hand side of the equation by making use of the Maxwell relationship that when $\mathrm{d}z = M\mathrm{d}x + N\mathrm{d}y$, and d$z$ is a total differential, then $\partial M/\partial y = \partial N/\partial x$. When we apply this relationship to our problem, we obtain: $(\partial S/\partial \omega)_T = \partial \varepsilon/\partial T$. In integrated form:

$$S = S_0 + \omega \frac{\mathrm{d}\varepsilon}{\mathrm{d}T}$$

at constant temperature and volume. The integration constant S_0 is the entropy with no field present.

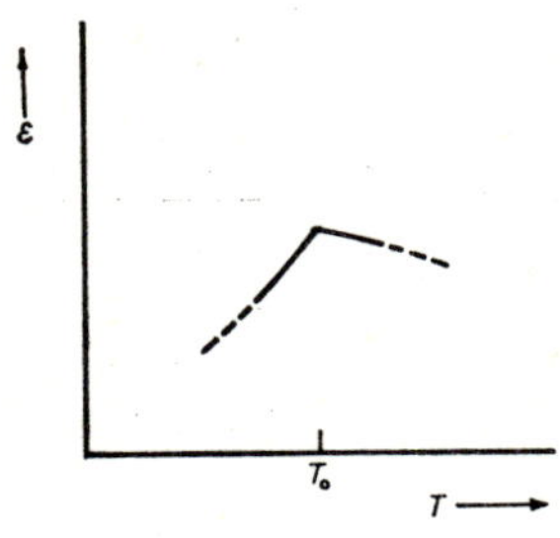

Fig. 2

An interpretation of this equation is that the effect of an applied field on the entropy of a system, at a given temperature, depends on the value of $\mathrm{d}\varepsilon/\mathrm{d}T$ at that temperature. For some solids, the experimental temperature dependence of dielectric constant is illustrated by the graph in Fig. 2. Below T_0, $(\mathrm{d}\varepsilon/\mathrm{d}T)$ is positive, and so according to our equation, an applied field would cause an entropy increase; in accordance with the Boltzmann interpretation of entropy as a measure of probability, this means that the application of a field would result in a less ordered system. Above T_0, $(\mathrm{d}\varepsilon/\mathrm{d}T)$ is negative, and so an applied field would increase the order of the system. One qualitative explanation of the experimental results which is consistent with these entropy considerations is as follows. Suppose that the material is a crystalline solid containing permanent dipoles. At very low temperatures the ordered crystal structure is that of dipoles oriented in some regular pattern so that there is no resultant total moment. Up to the temperature T_0, the structure tends to preserve itself against vibrations of the lattice; an applied field tends to align dipoles parallel to it, and any such alignment could cause only a decrease in the original order or an increase in the entropy. Above the temperature T_0, which may be the melting point, the dipoles are oriented at random. Then any alignment of the dipoles parallel to the field could only increase order or decrease entropy. A quantitative discussion of the alignment of permanent dipoles in the gas phase will be given later.

The expression for entropy, given above, can be incorporated into our equation for $\mathrm{d}F$ to yield expressions for free energy F and internal energy U. We begin with our expression for $\mathrm{d}F$, here separated into three terms:

$$\mathrm{d}F = \left(-S + \omega\frac{\mathrm{d}\varepsilon}{\mathrm{d}T}\right)\mathrm{d}T + \omega\mathrm{d}\varepsilon + \varepsilon\mathrm{d}\omega\,.$$

With this we use the entropy expression, solved for $-S_0$, which is:

$$-S_0 = -S + \omega\frac{\mathrm{d}\varepsilon}{\mathrm{d}T}.$$

Thus we can write:

$$\mathrm{d}F = -S_0\,\mathrm{d}T + \mathrm{d}(\omega\varepsilon)$$

when volume is constant. Integrating, we obtain:

$$F = F_0 + \omega\varepsilon\,,$$

where F_0 is the free energy when there is no external field, at a given temperature and volume.

The internal energy is most easily found from the definition $U = F + TS$. Simple substitution yields the expression:

$$U = U_0 + \omega\left(\varepsilon + T\frac{\mathrm{d}\varepsilon}{\mathrm{d}T}\right),$$

References p. 39

where U_0 is $F_0 + TS_0$, the internal energy at a given temperature and volume with no field applied.

From these equations it is apparent that $\omega\varepsilon$ or $\varepsilon E^2/8\pi$ is the change in free energy of a material substance (at constant temperature and volume) resulting from the application of a field. But $\omega\varepsilon$ does not represent the total change in internal energy, for in the equation for U there is the additional term $\omega T(\mathrm{d}\varepsilon/\mathrm{d}T)$ which represents change in the 'hidden' or thermal energy of the material upon application of the field. Only if the dielectric constant is independent of temperature (*i.e.*, if $\mathrm{d}\varepsilon/\mathrm{d}T = 0$) can the quantity $\omega\varepsilon$ be said to represent the change in internal energy with application of a field. However, it has been found that in many cases in which there is a temperature variation of dielectric constant, a term $A\omega$ represents the change of internal energy due to a field. This conclusion is based on the results of experiments which show that in many cases ε can by represented by $A + B/T$, where A and B are temperature independent*. Substituting this value into the equation for internal energy, we find $U = U_0 + A\omega$: in these cases the temperature-varying term B/T does not contribute to internal energy.

The Relation of Dielectric Constant to Polarizability of Orientation and of Deformation

We have found a relationship between the measureable quantity, dielectric constant, and polarizability per molecule. But a molecule's polarizability in an electric field is determined by two effects: (a) orientation of the molecule because of any permanent dipole moment it may have, and (b) the creation of an induced dipole which is a result of deformation of charge distribution by the field. In order to gain knowledge about molecular structure of a material, we need information about both these types of polarizability.

As a first step, we shall develop an expression for orientation polarizability. We begin by obtaining an expression for potential energy at any point in a field in terms of the dipole moment of a contained molecule at that point.

The model we shall use for a neutral molecule is an assembly of charges such that $\sum_i q_i = 0$. The position of each charge q_i is defined in a

* See Reference 1 for a List of References.

Cartesian system x, y, and z by ξ_i, η_i, and ζ_i. A potential $\Phi(x,y,z)$ exists at every point *; at the position ξ_i, η_i, ζ_i, the potential is Φ_i, which may be expressed in terms of potential at the origin, Φ_0, by expansion in the usual manner. With only linear terms retained, this expression is:

$$\Phi_i = \Phi_0 + \xi_i \left(\frac{\partial \Phi}{\partial x}\right)_0 + \eta_i \left(\frac{\partial \Phi}{\partial y}\right)_0 + \zeta_i \left(\frac{\partial \Phi}{\partial z}\right)_0.$$

Total potential energy for the molecule is simply:

$$V = \sum_i q_i \Phi_i ,$$

or:

$$V = \sum_i q_i \left[\Phi_0 + \xi_i \left(\frac{\partial \Phi}{\partial x}\right)_0 + \eta_i \left(\frac{\partial \Phi}{\partial y}\right)_0 + \zeta_i \left(\frac{\partial \Phi}{\partial z}\right)_0 \right].$$

The partial derivatives of potential at any point are equivalent to the negatives of the components of the field at that point. And since the term $\Phi_0 \sum_i q_i$ equals zero, the equation becomes:

$$V = -(X \sum_i q_i \xi_i + Y \sum_i q_i \eta_i + Z \sum_i q_i \zeta_i) ,$$

where X, Y, and Z are the field components. (That the three sums are independent of the position of the origin can be seen by rewriting the equation in terms of new coordinates formed by adding constants to the old coordinates: the result is the appearance of additional terms of the form $c \sum_i q_i$, all of which equal zero.)

We identify the three sums in the equation as the three components of the total dipole moment of the molecule. Therefore we can express the potential energy in vector notation, as:

$$V = -\tilde{E} \cdot \tilde{m} = -E\mu \cos \vartheta ,$$

where μ is the magnitude of the dipole moment and ϑ is the angle between the field and the dipole direction. In the development of the expression for energy we neglected non-linear terms, which involve derivatives of the field strength, and so these equations are exact only when the field is constant. An idea of the order of magnitude of μ can be obtained from the

* It should be noted that Φ is an externally caused potential; the mutual interaction potential of the charges, although large, is not important since they are fixed in position with respect to one another and their mutual energy is not available.

relation $\mu = ed$. The unit charge e is 4.8×10^{-10} esu; if d, the distance between charges in the dipole, is about one Ångström (10^{-8} cm), then μ is of the order of 10^{-18} esu cm. This value, designated the debye unit, is used as the unit of measure for dipole moment.

We are now ready to consider the extent to which permanent dipoles in the gas phase are oriented by an applied electric field, for as we shall see, our expression for potential energy of a molecule in a field will enable us to calculate the pertinent quantity, which is average dipole moment in the direction of the field.

Because the dipoles are in a state of constant collision, the problem is too complicated for any but statistical methods. Methods of statistical mechanics will not be discussed until a later chapter, but for the present we can obtain the statistical relation we need by making use of a well-known formula that applies to an analogous situation. This is the 'barometric formula', and the analogous situation is the dependence of gas density on height and temperature. The equation is *:

$$n = n_0 \, \mathrm{e}^{-mgx/kT} = n_0 \, \mathrm{e}^{-V/kT} \, ,$$

where n is the gas density at height x, g is the gravitational constant, m is the mass per molecule, n_0 is a constant, and the quantity V/kT is the ratio of the gravitational potential energy of a molecule to its thermal energy. With modification to the more general case, this simple formula becomes the expression known as Boltzmann's law. Applied to our problem, it means that the number of molecules having a certain potential energy V in the electric field is an exponential function of the ratio of this energy to the thermal energy.

In the absence of an external field, a gas of polar molecules will have no preferred orientation; if $\mathrm{d}n$ represents the most probable number of molecules within a solid angle $\mathrm{d}\Omega$ †, then $\mathrm{d}n = c\mathrm{d}\Omega$. In the presence of a field, the potential energy of a dipole is determined by its spatial orientation; in a polar gas, molecules oriented in the same direction have the same potential energy. If direction is indicated by the solid angle element $\mathrm{d}\Omega$, then according to Boltzmann's law the most probable num-

* A discussion of this equation is given in the Appendix p. 38.

† The solid angle subtended by a cone is equal to the area of intersection of the cone with a unit sphere whose center is at the vertex of the cone. The total solid angle around a point is thus 4π.

ber of molecules in $\mathrm{d}\Omega$ is:

$$\mathrm{d}n = C\,\mathrm{e}^{-V/kT}\,\mathrm{d}\Omega\,.$$

The potential energy of orientation for one molecule we found to be $-\mu E\cos\vartheta$, where ϑ is the angle measured from the field direction. Thus our expression of Boltzmann's law becomes:

$$\mathrm{d}n = C\exp\left(\frac{\mu E}{kT}\cos\vartheta\right)\mathrm{d}\Omega\,,$$

or, letting $\mu E/kT$ be represented by β:

$$\mathrm{d}n = C\mathrm{e}^{\beta\cos\vartheta}\,\mathrm{d}\Omega\,.$$

We are now able to calculate the average moment per molecule, $\overline{\mu}$, which will have the same direction as the field. The quantity $\overline{\mu}$ is obtained by performing a weighted average of the projection of μ in the field direction, the weighting factor being the probability of the particular orientation that is being considered. In concise form:

$$\overline{\mu} = \frac{\int_s \mu\cos\vartheta\,\mathrm{e}^{\beta\cos\vartheta}\,\mathrm{d}\Omega}{\int_s \mathrm{e}^{\beta\cos\vartheta}\,\mathrm{d}\Omega}\,,$$

where $\beta = \mu E/kT$. The angle $\mathrm{d}\Omega$ can be expressed in terms of the angle ϑ, for as shown in Fig. 3, $\mathrm{d}\Omega = 2\pi\sin\vartheta\,\mathrm{d}\vartheta$. Integration is carried out over the surface of a unit sphere; the angle ϑ varies from 0 to π. To perform the

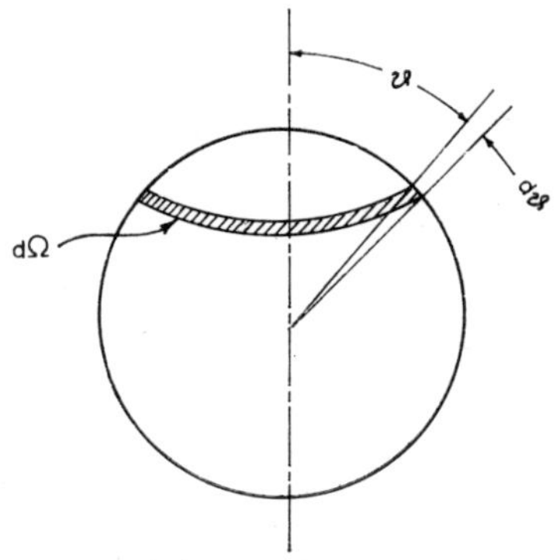

Fig. 3

References p. 39

integration, we let $x = \cos\vartheta$; then $dx = -\sin\vartheta\, d\vartheta$ and so $d\Omega = -2\pi\, dx$.

Rewriting the expression with this change of variable, we find:

$$\frac{\overline{\mu}}{\mu} = \frac{\int_{-1}^{+1} e^{\beta x}\, x\, dx}{\int_{-1}^{+1} e^{\beta x}\, dx}.$$

Integration yields the form:

$$\frac{\overline{\mu}}{\mu} = \frac{e^{\beta} + e^{-\beta}}{e^{\beta} - e^{-\beta}} - \frac{1}{\beta},$$

which is the same as $\coth\beta - 1/\beta$. An expression similar to this was first devised by Langevin [2] for the magnetic field, and is called the Langevin function. In terms of β this function is:

$$\mathscr{L}(\beta) = \coth\beta - (1/\beta).$$

The graphical appearance of this function is shown in Fig. 4. The limiting

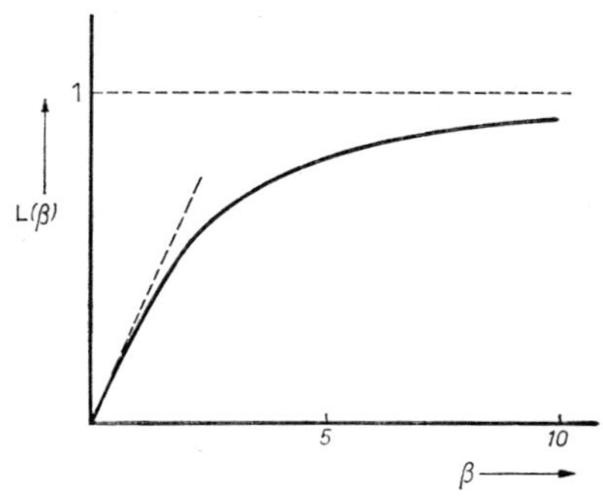

Fig. 4

value for $\overline{\mu}/\mu$ is, of course, unity. This is approached asymptotically as the field strength increases or as the temperature falls.

The order of magnitude of β is very small for ordinary cases: for example, if a field of one esu is applied, then since μ is about 10^{-18} esu cm and kT is about 4×10^{-14} ergs at room temperature, the value of $\mu E/kT$ is about 2.5×10^{-5}. Since the values of β being considered are so small, a Taylor expansion around $\beta = 0$ can be used, and this gives as a first approximation $\mathscr{L}(\beta) \approx \beta/3$. An even simpler way to obtain $\beta/3$ is to replace $e^{\beta x}$ with the approximation $1 + \beta x$, and then integrate. Summing up, we find that the average moment in the field direction is:

$$\bar{\mu} = \frac{\mu^2 E}{3kT}.$$

Orientation polarizability, α_0, can be obtained directly from $\bar{\mu}$, for by definition $\alpha_0 = \bar{\mu}/E$. Thus:

$$\alpha_0 = \mu^2/3kT.$$

The equation relating dielectric constant and polarizability per molecule can finally be written:

$$\varepsilon - 1 = 4\pi n\alpha = 4\pi n(\alpha_d + \alpha_0) = 4\pi n(\alpha_d + \mu^2/3kT),$$

where α_d is deformation polarizability, which results from deformation of the charge distribution of the molecule and is largely independent of temperature.

This temperature independence can be illustrated by consideration of a vibrating molecule like HCl. If Hooke's law is obeyed – that is, if the force between the masses is proportional throughout the range of vibration to the displacement from a given rest position – then the average separation between the masses is independent of vibrational energy. Therefore, if an external field were applied, the increase in average separation distance would be proportional to the field strength but independent of vibrational energy and hence of temperature. (This point is elucidated in the Appendix.) Another way of describing this property is to say: if the restoring force is proportional to displacement, then the potential energy of vibration will be proportional to the square of the displacement regardless of the direction of displacement. As long as this symmetry is preserved over the range of vibration, a change in the average separation distance during vibration is a function only of the strength of the external field. In terms of a polarizable particle, this means that deformation polarizability is temperature independent if the vibrational potential energy is symmetrical with respect to distance. At the temperatures generally used in the laboratory, most molecules vibrate in quite symmetrical ranges; the effect of dissymmetry has been calculated and found to be almost negligible. From these considerations we conclude

References p. 39

that it is reasonable to represent dielectric constant in the form $A + B/T$, both the constants being independent of temperature, with A arising from deformation and B arising from orientation of the permanent dipole.

We can write our equation in a more useful form involving quantities of a more convenient order of magnitude. The value of α is very small; for example, we found that for a polarizable conducting sphere of diameter a, the deformation polarizability equals a^3, which for a diameter of molecular size is of the order of 10^{-24} cm^3. But if we multiply α by Avogadro's number, N, which is of the order of 10^{24}, we obtain a quantity with the convenient order of magnitude of 1 cm^3; this quantity is called molar polarization and is defined as $P' = N\alpha$. The quantity n (particles per cm^3) may be written ρ/m, where ρ is density and m is particle weight in grams; then

$$n = \frac{\rho}{m} \cdot \frac{N}{N} = \frac{\rho N}{M},$$

where M is gram molecular weight. Accordingly:

$$\frac{\varepsilon - 1}{4\pi} \cdot \frac{M}{\rho} = N\left(\alpha_d + \frac{\mu^2}{3kT}\right) = P',$$

or, if for convenience both sides are multiplied by T:

$$\frac{\varepsilon - 1}{4\pi} \frac{MT}{\rho} = N\alpha_d T + N\frac{\mu^2}{3k}.$$

Experimentally, dielectric constant can be measured at various temperatures, and then the quantity

$$\frac{\varepsilon - 1}{4\pi} \frac{MT}{\rho}$$

plotted as a function of temperature: Fig. 5 shows a typical plot. Such a graph provides a means to evaluate the two types of polarizability, for according to our theory the slope gives $N\alpha_d$ and the intercept gives $N\mu^2/3k$ †.

† For typical data, see Reference 1 (p. 39).

Generally, dielectric constant is determined by bridge methods using alternating fields rather than by static methods which are very difficult. Orientation equilibrium is established very rapidly in the gas phase at ordinary temperatures, and so at sufficiently low frequencies the value of ε obtained is very close to its value when measured in a static field. At very high frequencies, however, the dipoles are unable to follow the field and the dielectric constant measured is due to deformation only. Deformation polarizability can also be calculated from measurements of optical refractive index, for according to a formula derived by Maxwell, dielectric constant (which at optical frequencies arises from deformation polarization only)* equals the square of refractive index. It is apparent that combinations of these various measurements can provide useful information about molecular structure. For example, if dielectric constant

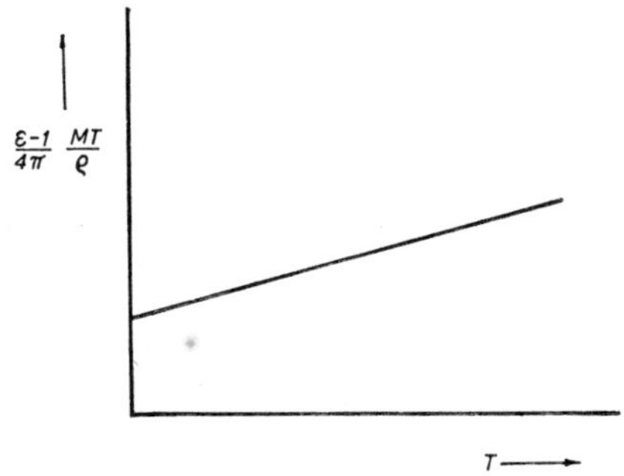

Fig. 5

measured at low frequency is much greater than the square of refractive index, it can be concluded that the molecules are certainly polar; in fact, the difference between the two measurements gives an evaluation of orientation polarizability and thus of permanent dipole moment.

We may make a final comment on the expression for orientation

* The deformation at very high frequency can be differentiated into two types. At visible frequencies there is present polarization due to the distortion of electron position in the molecule; at lower frequencies, in the infra red, there is, in addition, polarization of vibrating masses. It is found that for most materials the amount of polarization changes vary little over the frequency range of infra red to visible, but for ionic crystals there is a significant decrease. In the case of an ionic crystal, then, one should make the refractive index measurement in the infra red in order to detect all the polarization other than that due to orientation.

References p. 39

polarizability. With the help of the more recent quantum theory, it has been calculated that this term should be written $[1 - f(T)]\, \mu^2/3kT$. The correctional factor $f(T)$, according to Van Vleck [3], equals:

$$\frac{h^2}{48\pi^2 kT}\left\{\frac{\mu^2(x)}{\mu^2}\left[\frac{1}{B}+\frac{1}{C}\right]+\frac{\mu^2(y)}{\mu^2}\left[\frac{1}{A}+\frac{1}{C}\right]+\frac{\mu^2(z)}{\mu^2}\left[\frac{1}{A}+\frac{1}{B}\right]\right\}$$

where A, B, and C stand for the moments of inertia about the principal axes and $\mu(x)$, $\mu(y)$, and $\mu(z)$ are the projections of the dipole moment of the molecule onto the principal axes. However, the order of magnitude of $f(T)$ is, for most purposes, negligible. An appreciable correction has been calculated only in the case of HF, for which the moments of inertia are small. For HF, the correction term $f(T)$ changes the calculated moment from 1.91 debyes to 1.94 debyes.

Permanent Dipole Moment

Suppose an investigator has obtained an evaluation of permanent dipole moment on the basis of experimental measurements: of what use is this information? We can get an idea of the variety of possible applications by reviewing some experiments performed in past years.

Carbon dioxide is a molecule containing two different kinds of atoms, one of which attracts electrons to a greater extent than the other. For such a molecule, one might expect a separation between the positive and negative centers of charge. Measurement shows, however, that CO_2 has no permanent dipole moment, which means that it must be a linear molecule having the configuration represented by O=C=O. This has been supported by more recent spectroscopic measurements.

Non-zero dipole moments have, however, been measured for water and ammonia: the indication is that these molecules are not symmetrical. Again, this has been supported by independent methods, including spectroscopy, which indicate that ammonia is pyramidal and that the water molecule is non-linear.

Comparison of dipole moment measurements is an effective way to distinguish the three forms of disubstituted benzene. Like benzene itself, one of the three dichlorobenzenes is found to have a dipole moment of zero; the other two have different moments. By using the simple and fairly accurate principle that the total moment of a molecule is the vector sum of constituent moments (assumed independent of one another), one concludes that paradichlorobenzene has the zero moment, metadichlorobenzene has the lower non-zero moment, and orthodichlorobenzene has the highest moment.

A demonstration of the rigidity of double bonds is given by dipole moment studies of dichlorosubstituted ethylene. If the molecule were free to rotate about its double bond, there would be just two forms, a 1,1-dichloro- and a 1,2-dichloroethylene, both of which would have a non-zero moment, the first being much larger than the second. If, however, the double bond were rigid, there would be three distinct forms, all of which would have different dipole moments, including one which would be zero. Measurements do, in fact, indicate that this is the case. The 1,1-dichlorobenzene, which has the least symmetry, has the highest moment; the *cis* 1,2 form is intermediate, and the *trans* 1,2 form has a zero moment. The conclusion is that all the atoms are in the same plane and that there is no rotation about the double bond.

An important aspect of the molecular structure of certain organic compounds is demonstrated by dipole moment measurements at various temperatures. Let us consider the 1,2 disubstituted form of dichloroethane: for this we would expect to be able to write, as we have for other compounds, the temperature dependence of dielectric constant as $\varepsilon = A + B/T$, where B is a constant involving the dipole moment. However, in this case it is found that dielectric constant is a much more complicated function of temperature – the plot of dielectric constant times temperature versus temperature is not a straight line as it is for most compounds. This means that the moment μ is not constant but is, rather, a function of temperature: in the expression for ε, the term B (which includes μ^2) depends implicitly on T. We can explain this anomaly if we introduce the concept of 'hindered rotation', according to which the bulkiness of the atoms prevents the molecule from rotating freely about its single bond, the ease of rotation depending on the temperature. The concept of hindered rotation has been successfully used to explain many of the apparently paradoxical reactions of organic chemistry.

An effect that is of great interest in low-temperature experimentation is adiabatic demagnetization, which has made it possible to reach a temperature as low as 0.001^0K. A similar effect is to be expected when an electric field is removed adiabatically from a gas of permanent dipoles, and the magnitude of resulting temperature change can be calculated.

References p. 39

In the cooling process involving adiabatic demagnetization, a paramagnetic substance is first cooled by the Linde process. This is a cyclic process which produces cooling by free expansion of the gas (for the process to work, the gas must be already cooled below its inversion temperature, which is the point on the temperature scale above which free expansion heats a gas and below which it cools the gas). Quite low temperatures can be reached by this means; for example, helium, when liquified by the Linde process, can be cooled to 4^0 K, and when pumped to hasten evaporation, can be made to reach 1^0 K. When the cooled gas is a paramagnetic substance, a further lowering of temperature is achieved by the demagnetization process: a magnetic field is applied during the preliminary cooling, and then the field is adiabatically removed.

The effect of adiabatic removal of an electric field is calculated through thermodynamic considerations. As discussed previously, a polar gas, when placed in an electric field with temperature and volume kept constant, suffers a decrease in entropy. If then the field is removed adiabatically, the entropy must remain unchanged, in accordance with the thermodynamics of reversible processes. When the constraint of the field is removed, however, the molecules become more randomly oriented, and the entropy increase which would result must be compensated for by an entropy decrease realized through a fall in temperature. Essentially the calculation is as follows: under the influence of the field at temperature T_0, the initial entropy $S(T_0)$ is $S_0(T_0) + \omega(\mathrm{d}\varepsilon/\mathrm{d}T)$, as discussed earlier. When the field is removed adiabatically and reversibly, the new entropy $S(T_1)$ is given by $S_0(T_0) + (\partial S/\partial T)\, \Delta T$. Since these two values for entropy must be equal, and since $T(\partial S/\partial T) = \delta Q/\delta T \equiv C$, where Q is the heat put into the system and C is the heat capacity at constant volume, it is possible to write: $(C/T)\, \Delta T = \omega \mathrm{d}\varepsilon/\mathrm{d}T$. The value of $\mathrm{d}\varepsilon/\mathrm{d}T$, obtained by differentiating the expression $\varepsilon - 1 = 4\pi n(\alpha_d + \mu^2/3kT)$, is substituted into the previous expression and the resulting equation solved for ΔT:

$$\Delta T = -4\pi \frac{\mu^2}{3kT} \omega \frac{1}{\mathrm{C}/n} .$$

The magnitude of ΔT at room temperature is extremely small; for instance, if the applied field strength is 10 esu (3000 volts per cm), ΔT is calculated to be of the order of 10^{-5} ^{0}K. (Dipole moment μ is about 10^{-18} esu cm; energy density in vacuum, ω, is $E^2/8\pi$, which is about 4 ergs per cm^3; and the ratio C/n, which is heat capacity per molecule, is of the order of k, 10^{-16} ergs per degree.) At lower temperatures we calculate the effect to be larger since T appears in the denominator, but we cannot set T too low because this expression remains valid only while the polar material is in the form of a gas. Thus the effect on the temperature of a polar gas of adiabatic removal of the electric field is very small.

(*i*) *Derivation of the Differential Equation* curl $\tilde{E} = 0$

The experimentally found law that the total work involved in moving a charge around a closed path in an electric field is zero, which we have expressed in integral form, can also be expressed in differential form. The procedure is to take a starting point and write expressions for the work required to move a charge from this point around vanishingly small closed paths in each of three mutually perpendicular directions.

Let us begin with the xy plane, and choose as our path the boundaries of a small rectangle of dimensions Δx and Δy (see Fig. 6). The origin is

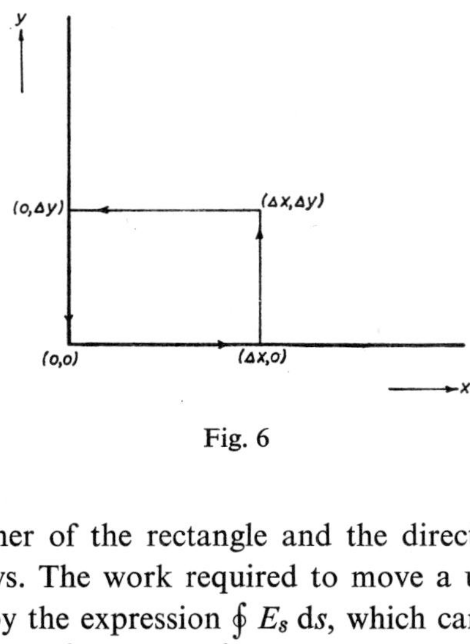

Fig. 6

placed at one corner of the rectangle and the direction of the path is indicated by arrows. The work required to move a unit charge around the path is given by the expression $\oint E_s \, ds$, which can also be written $\int_a E_x \, dx + \int_b E_y \, dy + \int_c E_x \, dx + \int_d E_y \, dy$; or more simply as $\int_a X \, dx + \int_b Y \, dy + \int_c X \, dx + \int_d Y \, dy$, where X is the x component and Y is the y component of the field E.

So far everything has been mathematically precise, but now for the practical purpose of evaluating these expressions, an assumption will be made. We shall assume that because the rectangle is considered to be very small, only linear terms in a Taylor expansion of the components of the field about the origin need be retained. Accordingly:

$$X(x, y) = X(0, 0) + \left.\frac{\partial X}{\partial x}\right|_{0,\,0} x + \left.\frac{\partial X}{\partial y}\right|_{0,\,0} y\,,$$

or, in shorter form, $X = ax + by + c$; similarly, $Y = ex + fy + g$. Here the constants a, b, e, and f represent the partial derivatives evaluated at the origin, and the ranges of x and y are given by $0 \leqslant x \leqslant \Delta x$ and $0 \leqslant y \leqslant \Delta y$. By simply substituting these expressions for X and Y back into the expression for $\oint E_s \,\mathrm{d}s$ and performing the integrations over the paths considered, we find: $\oint E_s \,\mathrm{d}s = -\, b\, \Delta y\, \Delta x + e\, \Delta x\, \Delta y$. Now in order to eliminate the arbitrary lengths Δx and Δy, we divide both sides of the equation by $(\Delta x\, \Delta y)$ and call the limiting case when Δx and Δy both approach zero, the curl of the field in the z direction. Since the quantities b and e are the partial derivatives $(\partial X/\partial y)$ and $(\partial Y/\partial x)$, the z component of curl $\tilde{E}$ is $(\partial Y/\partial x) - (\partial X/\partial y)$. The x and y components of curl $\tilde{E}$ are found in an analogous way. Since we know from our experimental law that each of these quantities must equal zero, we conclude that curl $\tilde{E} = 0$. In summary:

$$\frac{\partial X}{\partial z} - \frac{\partial Z}{\partial x} = 0\,; \frac{\partial Y}{\partial x} - \frac{\partial X}{\partial y} = 0\,; \frac{\partial Z}{\partial y} - \frac{\partial Y}{\partial z} = 0\,;$$

or:

$$\operatorname{curl} \tilde{E} = 0\,.$$

That these expressions are independent of the choice of position of the origin may be verified by choosing an arbitrary point x_0, y_0, z_0, and carrying through the above line of reasoning; identical results will follow.

(*ii*) *Demonstration that the Definition of Potential is Consistent with the Experimental Law*, $\oint E_s \,\mathrm{d}s = 0$

In accordance with the definition of potential, the field components X, Y, and Z can be written as $-\,\partial\Phi/\partial x$, $-\,\partial\Phi/\partial y$, and $-\,\partial\Phi/\partial z$. The quantity E_s, the projection of the field in the direction of a segment $\mathrm{d}s$, is the sum of the projections of the field components: $E_s = X\cos\alpha + Y\cos\beta + Z\cos\gamma$,

References p. 39

where α, β, and γ are the angles between the three axes x, y, and z, and the segment ds. But the cosines can also be written dx/ds, dy/ds, and dz/ds, so that E_s equals $-(\partial\Phi/\partial x)(\mathrm{d}x/\mathrm{d}s) - (\partial\Phi/\partial y)(\mathrm{d}y/\mathrm{d}s) - (\partial\Phi/\partial z)(\mathrm{d}z/\mathrm{d}s)$, which is simply $-\mathrm{d}\Phi/\mathrm{d}s$. Therefore, the circuit integral $\oint E_s\,\mathrm{d}s$ is

$$-\int_A^A \mathrm{d}\Phi\,,$$

which vanishes identically. This is consistent with our 'first experimental law', which we express as $\oint E_s\,\mathrm{d}s = 0$.

(iii) Derivation of the Differential Equation, div $\tilde{E} = 4\pi\rho$

We have expressed our 'second experimental law' in integral form as $\int_s E_n\,\mathrm{d}\sigma = 4\pi Q$, where the integration is performed over the entire surface enclosing a volume containing the total charge Q. As we have discussed, the quantity $4\pi Q$ can be thought of as the number of lines of force passing outwards through the enclosing surface. Now we wish to derive a differential expression of this law.

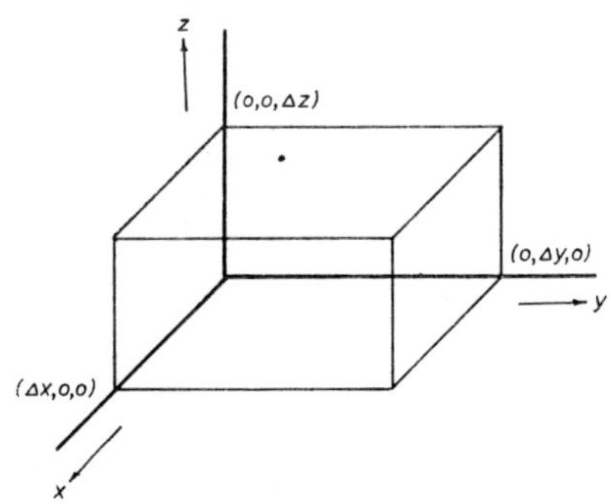

Fig. 7

We begin by finding the number of lines of force that originate within a small volume in a field. We consider for this purpose a very small parallelepiped with sides Δx, Δy, and Δz enclosing a charge distribution, and put the origin of coordinates at one corner (see Fig. 7). Our procedure will be to find the difference between the number of lines of force entering and leaving each of the three pairs of parallel faces, and add these together to obtain total lines originating in the volume.

First of all, we know that density of lines of force at any point is defined as the field strength at that point, and since the direction of a line is the same as that of the field, the density of lines of force in each of the three directions is represented by the field components X, Y, and Z. By means of a Taylor expansion with only linear terms retained, we can write for the X, Y, and Z components:

$$X = p_1x + q_1y + r_1z + X_0 \;;$$
$$Y = p_2x + q_2y + r_2z + Y_0 \;;$$
$$Z = p_3x + q_3y + r_3z + Z_0 \;;$$

where the constants p, q, and r represent the partial derivatives of the field components evaluated at the origin, the final terms represent the field components at the origin, and x, y, and z lie within Δx, Δy, and Δz respectively.

Now let us consider the two faces in the yz plane. The contribution of these faces to the total number of lines of force originating within the volume is the number of lines leaving the front face, minus the number entering the rear face. Since the Y and Z components do not penetrate these faces, we need consider only the X component, which is $\int_{\Delta y \Delta z} X(\Delta x, y, z)\, dydz - \int_{\Delta y \Delta z} X(0, y, z)\, dydz$. Substituting for X the evaluations we obtained by Taylor expansion, combining integrands, and integrating, we obtain for our number the expression, $p_1\, \Delta x \Delta y \Delta z$

The contributions from the other two pairs of faces are calculated in the same way; by summing all these contributions, we find that the total number of lines of force originating in the volume element is $(p_1 + q_2 + r_3)$ $\Delta x \Delta y \Delta z$. Since we let the constants represent certain partial derivatives of the field, this quantity is actually $(\partial X/\partial x + \partial Y/\partial y + \partial Z/\partial z)\, \Delta x \Delta y \Delta z$.

According to experiment, the number of lines of force emanating from a volume equals 4π times the total enclosed charge. In our case, this quantity is $4\pi\rho\, \Delta x \Delta y \Delta z$, where ρ is charge per unit volume, and so $\partial X/\partial x + \partial Y/\partial y + \partial Z/\partial z = 4\pi\rho$. This sum of partial derivatives, representing the number of lines originating per unit volume, is termed the divergence of the field and is abbreviated as div $\tilde{E}$. Thus the final result is written:

$$\text{div}\, \tilde{E} = 4\pi\rho$$

for a vacuum distribution of charge. We note that this divergence operation yields a scalar quantity which is independent of the position of the origin.

References p. 39

(iv) *Calculation of Capacity for a Vacuum-Filled Condenser*

By experiment it is found that the charge Q accumulated by a parallel plate condenser is proportional to the potential difference, Φ_0, between the plates. This is expressed by the formula $Q = C\Phi_0$, where C is a proportionality constant called capacity, which we want to express in terms of the geometry of the condenser.

Between the condenser plates there is no charge, and so Laplace's equation applies. Because of the symmetry of the parallel plate condenser (aside from edge effects), we know that the field strength should be uniform on any plane between and parallel to the faces; this means that for this application Laplace's equation is written $\mathrm{d}^2\Phi/\mathrm{d}z^2 = 0$, where z is the coordinate perpendicular to the plates. Solving this simple equation we find that $\Phi = az + \text{constant}$; and calling the potential of the plate with the negative charge zero, we find that $\Phi = az$. The field strength is $-\,\mathrm{d}\Phi/\mathrm{d}z$ or $-\,a$, and so is seen to be a constant field directed from the positive to the negative plate. In terms of the total potential difference Φ_0, the magnitude of the field is Φ_0/d, where d is the distance between the plates.

We can now apply the experimental law which holds for the vacuum case, $\int E_n \mathrm{d}A = 4\pi Q$, to our condenser, choosing the surface A as shown

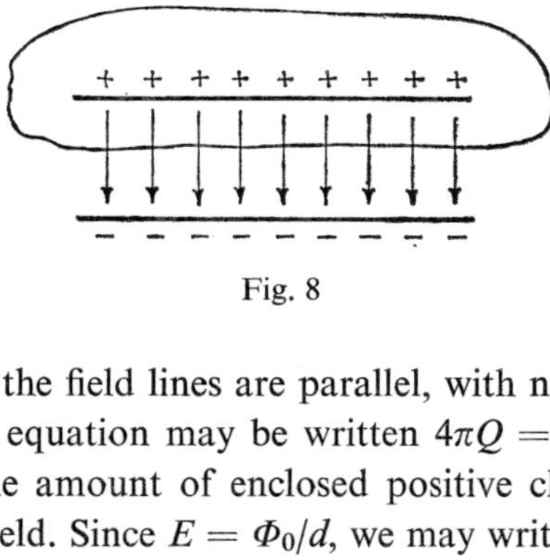

Fig. 8

in Fig. 8. Since all the field lines are parallel, with none existing outside the condenser, the equation may be written $4\pi Q = ES$, where S is the plate area, Q is the amount of enclosed positive charge, and E is the magnitude of the field. Since $E = \Phi_0/d$, we may write:

$$Q = \frac{1}{4\pi}\frac{S}{d}\Phi_0 \,.$$

Hence capacity C equals $S/4\pi d$ in vacuum.

(v) *Potential due to a Point Charge*

Absolute potential, in the usual sense, is defined as the work done by the system in moving a unit positive charge from a given point in a field to a point infinitely far removed where the potential is defined as zero. If the field arises from a point charge q, it is radially directed and has the magnitude q/r^2, where r is distance from q. The potential at a given point r_a equals the work the field would do in moving a unit positive charge from r_a out to infinity; since work is independent of path, the path can be chosen as radial, and so the work involved is simply

$$\int_{r_a}^{\infty} \frac{q}{r^2}\,\mathrm{d}r\,,$$

or

$$\frac{q}{r_a}\,.$$

The important point to be noticed is that this potential varies inversely as the first power of the radial distance and is spherically symmetrical.

(vi) *Potential due to a Dipole*

A dipole consists of two equal and opposite charges, $+q$ and $-q$, separated by a distance d. We consider a point P at a distance which is large compared to d, and choose a coordinate system with the origin at the charge $-q$ and the positive z axis in the direction of the charge $+q$ (see Fig. 9). The potential at P is the sum of potentials arising from the

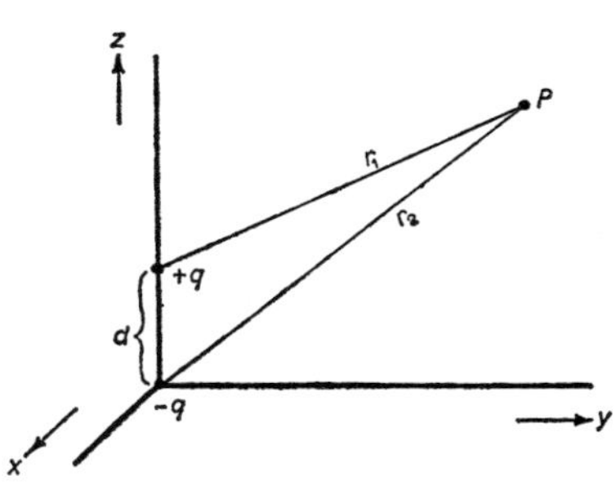

Fig. 9

References p. 39

two charges (this derives from the experimental law that the force, due to a collection of charges, acting on an exploratory charge is the sum of the forces due to the individual charges). The potential Φ_P at P is, then, $q/r_1 - q/r_2$. For very large values of distance between the dipole center and the point P, this potential may also be written:

$$\Phi_P = q\left[-\frac{\partial(1/r)}{\partial z}\right]d\,.$$

(This expression may be easily checked by writing out the definition of the derivative as the limiting case of a ratio of differences.) Performing the differentiation, we obtain $\Phi_P = (qd/r^2)(z/r)$. Since z/r is $\cos\vartheta$ and qd is dipole moment m, the formula is $\Phi_P = (m/r^2)\cos\vartheta$.

It will be recalled that potential due to a point charge was shown to vary with the inverse of the radial distance. Now we see that potential due to a dipole varies inversely with the square of distance, and has an angular dependence on dipole orientation. In similar manner, it is possible to show that potential resulting from higher multipoles varies inversely with higher powers of distance.

(*vii*) *The Electrical Equivalence in a Field of a Conducting Sphere and a Dipole*

It can be shown that, when viewed from the outside, a conducting sphere in a uniform field is equivalent electrically to a 'point dipole' of moment $m = a^3E$, where a is the radius of the sphere.

The term 'conducting sphere' implies that charges are free to move upon the surface. This means that there can be no work required to move a charge from one point to another on the surface, and therefore the potential on the surface is constant. The potential is, in fact, constant throughout the volume, since there can be no lines of force within an empty shell of constant potential. Suppose such a conducting sphere is placed in a vacuum under the influence of a homogeneous field (constant everywhere in magnitude and direction). We choose the origin of coordinates as the center of the sphere, and let this also be the zero point of potential. Since the potential is constant throughout the sphere, the surface potential must be zero.

One could now solve for the potential at any external point by employing Laplace's equation, $\Delta\Phi = 0$, with the added condition that the potential is to be zero on the surface of the sphere. This expression for potential would be identical to that describing the potential due to a 'point dipole', located at the original sphere center, except that in place of moment m there would appear the quantity a^3E.

A simpler way of arriving at the same result is to begin with the idea that a dipole of some undetermined moment is located at the center of a hypothetical sphere, and then calculate what moment the dipole would have to have in order to make the hypothetical sphere appear electrically like an actual conducting sphere. We must assume that the radius a of the sphere is large compared to the charge separation distance d of the dipole. We again locate the origin of coordinates at the center of the sphere and stipulate that this is the zero point of potential; also, we let the orientation of the dipole be with its positive charge in the positive direction of the z axis. At points on the 'surface' where z equals some particular value z_0, the potential due to the dipole only is $\Phi_0 = (m/a^2)(z_0/a)$ as shown in section (*vi*); the potential due to the undisturbed external field at the same points on the surface is $- z_0E$, as discussed in section (*iv*). Now if our hypothetical sphere is to appear like an actual conducting sphere, the potential at the 'surface' must equal zero. It must be, then, that $mz_0/a^3 - Ez_0 = 0$, or induced dipole moment m must equal a^3E.

The interpretation is that the charge distribution on the surface of a conducting sphere produces a reaction field which, to an outside observer, would be indistinguishable from that produced by an induced 'point' dipole of moment a^3E situated at the original location of the sphere center. Since polarizability is defined as $\alpha = m/E$, for this case α equals a^3, and so is seen to be proportional to the sphere volume.

(*viii*) *The Similarity between an Atom and a Metal Sphere*

Consider the Bohr model of the hydrogen atom: a central positive nucleus surrounded by a circular electron orbit (see Fig. 10). Suppose in an external field, the electron is displaced a distance l to another orbit whose plane is parallel to the first (the field being perpendicular to the orbit plane). Let r be the distance from the nucleus to the orbit. The restoring force in the vertical direction will be the vertical projection of the mutual attraction between the nucleus $(+ e)$ and the electron $(- e)$, namely $(e^2/r^2)(l/r)$. This must be balanced by the force on the electron

References p. 39

due to the presence of the field E; that is, by eE. Equating the two forces, we find that at equilibrium

$$eE = \frac{e^2 l}{r^3};$$

and since el is defined as moment m, the final result is $m = el = r^3 E$. The polarizability α is thus r^3, the cube of the atomic radius. This result is the same as that calculated (in section (*vii*)) for polarizability of a conducting sphere, and so the conclusion is that an atom of this model would behave electrically like a metal sphere.

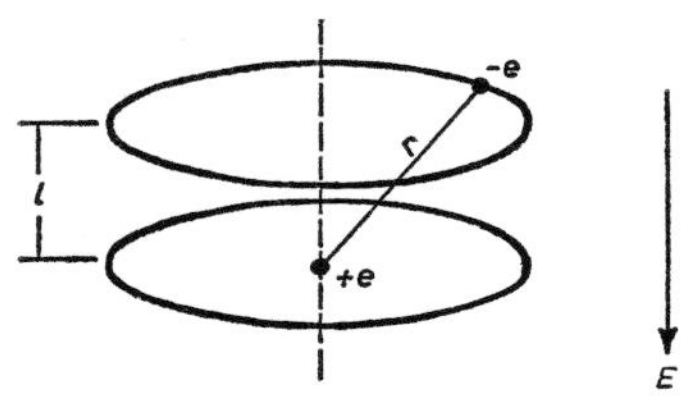

Fig. 10

(*ix*) *Derivation of Electric Displacement in Terms of an Equivalent Vacuum Distribution of Charge*

It has been shown that the divergence of field strength in a vacuum is proportional to the charge density there: div $\tilde{E} = 4\pi\rho$. We shall now consider how this relationship is modified by the presence of polarizable material, and derive a new expression applicable to dielectric.

If we represent a dielectric body by a vacuum charge distribution occupying the same volume and having the same electrical effect, then the volume containing the body is effectively a vacuum and the vacuum law can be applied. In place of free charge density, ρ, in the equation, there will be a term $(\rho + \rho^*)$ where ρ^* represents the vacuum charge density whose effect is the same as that of the polarized material. The problem is to express ρ^* in terms of some parameter describing the state of the dielectric: we shall use for this parameter the dipole moment per unit volume, P. Our procedure will be to calculate the potential, at some arbitrary outside point, that is produced by a volume distribution of material having a dipole moment per unit volume, P. Since the quantity ρ^* is the charge density which would produce this same potential if the volume under consideration were a vacuum, the two expressions for potential are equal, and ρ^* can thus be expressed in terms of P.

The potential Φ produced at some point by a single dipole of moment m has been shown to equal $(m/r^2)\cos\vartheta$. In vector notation, $\Phi = -\tilde{m}\cdot\tilde{r}/r^3$, where the vector $\tilde{r}$ extends from the point at which the potential is to be determined to the dipole. If instead of a single dipole we have a volume enclosing a distribution of dipoles with a macroscopic dipole moment per unit volume P, we can consider each volume element $\mathrm{d}v$ as a source of dipole moment $P\mathrm{d}v$. Each volume element produces a potential of $-(\tilde{P}\cdot\tilde{r}/r^3)\,\mathrm{d}v$ at the outside point, and total potential Φ_T is obtained by integrating over the volume:

$$\Phi_T = -\int_V \frac{\tilde{P}\cdot\tilde{r}}{r^3}\,\mathrm{d}v\,.$$

We assume with no loss of generality, that P falls off rapidly but smoothly to zero at the boundary. The integrand can be written in terms of the divergence of polarization, as can be seen by considering the quantity $\operatorname{div}(\tilde{P}/r)$. In accordance with the definition of divergence, we may write:

$$\begin{aligned}\operatorname{div}\frac{\tilde{P}}{r} &= \frac{\partial}{\partial x}\frac{P_x}{r} + \frac{\partial}{\partial y}\frac{P_y}{r} + \frac{\partial}{\partial z}\frac{P_z}{r}\\ &= P_x\frac{\partial(1/r)}{\partial x} + P_y\frac{\partial(1/r)}{\partial y} + P_z\frac{\partial(1/r)}{\partial z} + \frac{1}{r}\left(\frac{\partial P_x}{\partial x} + \frac{\partial P_y}{\partial y} + \frac{\partial P_z}{\partial z}\right)\\ &= -P_x\frac{x}{r^3} - P_y\frac{y}{r^3} - P_z\frac{z}{r^3} + \frac{1}{r}\left(\frac{\partial P_x}{\partial x} + \frac{\partial P_y}{\partial y} + \frac{\partial P_z}{\partial z}\right),\end{aligned}$$

which in abbreviated form is:

$$\operatorname{div}\frac{\tilde{P}}{r} = -\frac{\tilde{P}\cdot\tilde{r}}{r^3} + \frac{1}{r}\operatorname{div}\tilde{P}\,.$$

If this expression is solved for $\tilde{P}\cdot\tilde{r}/r^3$ and the result substituted into the potential expression, one finds:

$$\Phi_T = \int_V \operatorname{div}\frac{\tilde{P}}{r}\,\mathrm{d}v + \int_V -\frac{\operatorname{div}\tilde{P}}{r}\,\mathrm{d}v\,.$$

References p. 39

We know from a theorem of vector analysis that the volume integral of a divergence is equivalent to the integral of the normal component of the vector over the enclosing surface. Thus the first volume integral in the expression for Φ_T can be written:

$$\int_\sigma (P_n/r)\,\mathrm{d}\sigma\,,$$

where σ is the surface enclosing the volume V. Since the surface was drawn so as to completely enclose all the dipolar material, and since P falls rapidly but continuously to zero at the boundary, the dipole moment at the surface is zero and the integral vanishes. The second integral, then, is the expression for potential at the given outside point due to the polarization of dipolar material.

The potential at the outside point which would arise from an equal volume enclosing the vacuum charge density, ρ^*, is:

$$\Phi_T = \int_V \frac{\rho^*}{r}\,\mathrm{d}v\,.$$

Equating the two expressions for total potential we obtain:

$$\int_V -\frac{\mathrm{div}\,\tilde{P}}{r}\,\mathrm{d}v = \int_V \frac{\rho^*}{r}\,\mathrm{d}v, \quad \text{or } \rho^* = -\,\mathrm{div}\,\tilde{P}\,.$$

As we have indicated, the quantity $(\rho + \rho^*)$ must be substituted for ρ in the expression div $\tilde{E} = 4\pi\rho$ in order to make the equation applicable to dielectric:

$$\mathrm{div}\,\tilde{E} = 4\pi(\rho + \rho^*) = 4\pi\rho - 4\pi\,\mathrm{div}\,\tilde{P}\,.$$

Collecting terms, we obtain:

$$\mathrm{div}\,(\tilde{E} + 4\pi\tilde{P}) = 4\pi\rho\,.$$

The quantity $(\tilde{E} + 4\pi\tilde{P})$ is termed the electric displacement vector $\tilde{D}$, and so the new expression may be written:

$$\mathrm{div}\,\tilde{D} = 4\pi\rho\,,$$

where ρ is density of any free charge that may be present within the dielectric body.

(x) *A Vector Transformation of an Equation for Differential Change in Electrical Work*

In our discussion of energy density, we stated that the equation $\mathrm{d}W = \int_V \Phi\, \mathrm{d}\rho \mathrm{d}v$ can be transformed to:

$$\mathrm{d}W = \frac{1}{4\pi} \int_{\text{all space}} \tilde{E} \cdot \mathrm{d}\tilde{D}\, \mathrm{d}v\,.$$

We shall show this as a vector transformation.

If a vector $\tilde{B}$ is such that $\tilde{B} = u\tilde{A}$, where u is a scalar function of the coordinates, then one can write $\int_V \operatorname{div} \tilde{B}\, \mathrm{d}v = \int_V u \operatorname{div} \tilde{A}\, \mathrm{d}v + \int_V \tilde{A} \cdot \operatorname{grad} u\, \mathrm{d}v$, as can be verified by writing out each term. (The term $\tilde{A} \cdot \operatorname{grad} u$ is an abbreviation for $A_x(\partial u/\partial x) + A_y(\partial u/\partial y) + A_z(\partial u/\partial z)$.) Also, we know from a theorem of vector analysis that the volume integral $\int_V \operatorname{div} \tilde{B}\, \mathrm{d}v$ can be written as the surface integral $\int_S \tilde{B}_n\, \mathrm{d}\sigma$. If we identify the scalar quantity u as potential Φ, and $\tilde{A}$ as the vector $\mathrm{d}\tilde{D}/4\pi$, we obtain:

$$\int_V \Phi \operatorname{div} \frac{\mathrm{d}\tilde{D}}{4\pi}\, \mathrm{d}v + \int_V \frac{\mathrm{d}\tilde{D}}{4\pi} \cdot \operatorname{grad} \Phi\, \mathrm{d}v = \int_S \left(\Phi \frac{\mathrm{d}\tilde{D}}{4\pi} \right)_n \mathrm{d}\sigma\,.$$

The surface integral will vanish if the integration is performed over all space. No matter what the charge distribution is, at extremely far distances it will behave like a point charge; its potential will fall off as $1/r$, where r is the distance from the center of the charge distribution to the point considered. The change in displacement $\mathrm{d}\tilde{D}$ due to a change in the value of charge density at every point will fall off as $1/r^2$, and so the product of Φ and $\mathrm{d}\tilde{D}$ goes to zero as $1/r^3$. Since the area of the enclosing surface increases as r^2, the surface integral goes to zero as $1/r$, and so is evaluated as zero over all space.

Since $4\pi\rho = \operatorname{div} \tilde{D}$, we may write $\mathrm{d}\rho = \operatorname{div} \mathrm{d}\tilde{D}/4\pi$; also, the term $-\operatorname{grad} \Phi$ is the field vector $\tilde{E}$. Substituting these quantities into the volume integrals, we find:

$$\int_V \Phi\, \mathrm{d}\rho\, \mathrm{d}v = \frac{1}{4\pi} \int_{\text{all space}} \tilde{E} \cdot \mathrm{d}\tilde{D}\, \mathrm{d}v\,.$$

References p. 39

(*xi*) *The Barometric Formula, a Special Case of Boltzmann's Law*

Suppose we have a column of gas in which it is assumed that there are no interactions between the molecules. The equation of state for such a gas is $P = nkT$, where n represents the number of molecules per cubic centimeter. The problem is to find how n varies with height.

If the cross-section of the column is one square centimeter and the distance from the base to any point is x, then in a segment $\mathrm{d}x$, the gravitational force is $\rho g \,\mathrm{d}x$, where ρ is the gas density in grams per cubic centimeter, and g is the gravitational constant (see Fig. 11). The pressure

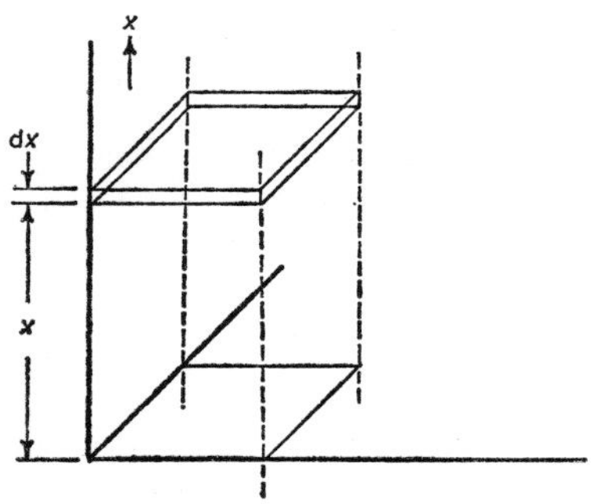

Fig. 11

difference from the bottom of the segment to the top is given by $(\mathrm{d}P/\mathrm{d}x)\,\mathrm{d}x$. At equilibrium the net force is zero, or $\mathrm{d}P/\mathrm{d}x + \rho g = 0$. Substituting for P the expression derived from the perfect gas law, and for ρ the quantity nm, where m is the mass of one molecule, we obtain:

$$kT(\mathrm{d}n/\mathrm{d}x) + nmg = 0\,.$$

Separating the variables and integrating, we find:

$$n = n_0 \mathrm{e}^{-mgx/kT}\,,$$

where n_0 is a constant and mgx is gravitational potential energy. This is a special case of Boltzmann's law.

(*xii*) *Shift in Average Position of a Vibrating Molecule under the Influence of a Constant External Force*

Let us visualize a vibrating molecule as a mass attached to a spring. Suppose the mass is oscillating and at a certain instant a uniform attracting force is applied (for example, the mass may be charged and a homogeneous field turned on). If Hooke's law is obeyed, the change in the average position of the mass will be a function of the field strength only – the amplitude of vibration of the mass will have no influence. This is shown by consideration of the applicable equations of motion. When there is no applied field, the equation of motion of the mass is $\ddot{x}_1 + \omega_0{}^2 x_1 = 0$, where x_1 is the displacement from rest position, and $\omega_0{}^2$ represents the square of the angular frequency of vibration and equals k/m, the spring constant divided by the mass of the particle. The solution for x_1 in terms of time is $x_1 = A \cos(\omega_0 t + \varphi)$, where A is the amplitude of vibration and φ is an arbitrary phase angle. When a field is applied, the new equation of motion is $\ddot{x}_2 + \omega_0{}^2 x_2 - S = 0$. Here S represents the acceleration due to the field. Solution is facilitated by the substitution $y = x_2 - S/\omega_0{}^2$; we obtain, after integration, the relation $x_2 = B \cos(\omega_0 t + \varphi') + S/\omega_0{}^2$.

We know that the average value of the cosine of an angle is zero, and so the average value of x_1 is zero. The average value of x_2, however, is $S/\omega_0{}^2$, a quantity that depends only on the field and not on the original vibrational energy which is a function of the amplitude A.

REFERENCES

1. Debye, P., *Polar Molecules*, Dover, New York, 1929, Chapter III.
2. Langevin, P., *J. phys.*, [4] 4 (1905) 678; *Ann. chim. et phys.*, [8] 5 (1905) 70.
3. Van Vleck, J. H., *Electric and Magnetic Susceptibilities*, Oxford Univ. Press, 1932, p. 198.

CHAPTER 2

The Statistical Method

A statistical method is required for the theoretical description of many physical situations. We have, in fact, already used a statistical procedure in our treatment of dipole orientation. However, we have not yet presented a sound basis for statistical method; in our treatment of the dipole orientation problem we obtained our method merely by inference from an analogous situation which is described by the well known and experimentally proven 'barometric formula'. In this chapter we shall discuss the theoretical basis of statistical mechanics. Further, we shall show the development of statistical methods for the calculation of thermodynamic functions, and examine their validity by comparison with experimental measurements.

The fact that a statistical method is indispensable for the calculation of the mechanics of large numbers of particles is quite apparent. A problem involving only two interacting particles can be solved by methods of mechanics. But the mechanics of even three interacting particles (the famous three-body problem) is so complicated that no general solution can be obtained; in some cases an approximate solution can be obtained, but only with the use of laborious procedures. For larger groups of particles, such as the 6×10^{23} molecules in a mole of material, it is obviously impossible to obtain a solution even by approximation procedure.

But although we leave behind the ability to determine individual behavior as we consider larger and larger numbers of particles, we enter a realm where we may predict with increasing accuracy the fraction of particles having a given property: in other words, we may predict the probability of occurrence of certain properties or events. The method is called statistical mechanics and its fundamental postulate is that the behavior of a very great number of identical interacting particles or systems can be predicted mathematically by means of methods of statistics, the accuracy of a prediction increasing with increasing number of particles or systems.

How do we know that the principle of probability is reliable? Certainly it cannot be tested through calculation of the mechanics of each of a tremendous number of particles. The reasonableness of the probability principle is demonstrated in everyday experiences such as the playing of games of chance (tossing of coins, rolling of dice, playing of cards, etc.), where the probability of a result may be accurately predicted. It was, in fact, partly because of games of chance that many years ago the science of probability and statistics came into being.

The earliest statistical method, the Maxwell-Boltzmann statistics, is based on the idea that a single particle such as an atom or a molecule comprises the system which is repeated N times in a macroscopic collection of N molecules. It is considered that the N systems are contained at constant volume, the container being kept isolated so that the total energy remains constant. The systems are assumed to interact enough to ensure equilibrium, yet so little that the probability for some chosen condition of the entire collection can be represented as the product of the probabilities for each molecular condition or state*. A most probable condition is calculated for the collection of systems, and the reasonable assumption is made that this most probable condition is that which occurs physically. Relative deviations are shown to approach zero as the number of molecules considered is increased.

Two fundamental deficiencies of this theory are readily apparent. (Another very important one became evident when quantum mechanics was introduced; we shall discuss this later.) The first lies in the fact that the number of systems, though tremendous, is limited. The second and more serious drawback is that the systems must be assumed to interact very little – just enough to provide a means for the establishment of dynamic equilibrium. If there were a great deal of interaction, the composite probabilities could not be calculated through individual probabilities alone, and the problem would get out of hand. An ideal gas, then, is the only collection of systems that can be treated in this manner.

* This is completely analogous to the problem of calculating probability in the tossing of coins. If N coins, c_1, c_2, . . . c_N, are tossed, the probability that the first will be, say, heads, the second, say, tails, etc., is just the product of the individual probabilities. In symbols one would write $P(c_1)\,P(c_2)\,\ldots\,P(c_N)$.

References p. 85

Years later, J. W. Gibbs formulated the problem in such a way that these two difficulties were not present. He took for his single system the entire container of N particles and considered it to be immersed in a heat bath made up of L identical containers of particles; L was considered to represent an unlimited number. In the theory it is assumed that each container can exchange energy through its walls with other containers, thus enabling the entire collection to reach equilibrium; yet the rate of energy exchanged is considered to be so small that the systems are virtually independent. The entire collection of systems is considered as isolated from the rest of the universe so that it has a constant energy. This 'Gibbs Ensemble', as it is called, has the following properties: (a) Because the systems are virtually independent, the composite probability for a given configuration of the ensemble is the product of system probabilities. (b) Since there is so little interaction between systems, and since interactions within the systems do not come into consideration, the material need not be an ideal gas: even liquids and solids can be considered. (c) Each system, being immersed in an infinite heat bath of constant energy, is maintained at a constant temperature. It is an advantage to be able to consider the system as at constant temperature rather than constant energy, because this corresponds to an experimentally feasible condition. (d) There is an unlimited number of systems.

In this theory, as in the Maxwell-Boltzmann theory, it is assumed that the values calculated to be most probable are the ones represented in nature. As is well discussed in Schroedinger's book[1], it is possible and really more rigorous to calculate properties through average rather than most probable values, but it is shown that because the number of identical systems is taken to be unlimited, the two methods yield identical results.

Statistical theory was first used in conjunction with the ideas of classical mechanics. But the modern way of applying statistical methods is to use the ideas of quantum mechanics as applied to the Gibbs model. Therefore we shall begin our mathematical development in terms of the Gibbs theory and the quantum mechanical approach.

Our Gibbs ensemble, then, consists of macroscopic systems which exist in discrete states having discrete energies. The assumption is made that these states are, *a priori*, equally probable. A configuration of the ensemble is described by specifying how many systems exist in each possible state: each of these numbers is expressed as the number, L_i, of systems in the state (whose associated energy is u_i). Using the basic principles of probability theory, with our assumption of the independence of systems, we can write an expression for the number of ways in which the systems can be arranged for any given configuration. In other words, we indicate how many different arrangements can be obtained by distributing the systems among the l available states while always keeping within the same configuration $L_1, L_2, \ldots L_l$. At this point we must clearly specify what we mean by 'different arrangements'. Since our Gibbs systems are all macroscopic containers and therefore distinguishable, we shall assume that a changing of places between systems in different states results in a different arrangement, whereas no new arrangement is produced by a reshuffling of systems within each state. The expression is:

$$P = \frac{L!}{L_1!\, L_2! \ldots}$$

where P is an unnormalized probability. The sum of L_i's must add up to the total number of systems, L, in the ensemble (here L is taken as large but finite, for ease of calculation; later limits will be taken as L becomes infinite). Also, the L_i's must comply with the fact that $\sum_i L_i u_i$ equals total energy U. The number P may be considered as a measure of disorder for the given configuration of the ensemble.

The Relation of Probability to Thermodynamic Properties; Partition Function

A way of relating probability to a macroscopic physical quantity was introduced by Boltzmann, who postulated that probability should be a measure of entropy. This is an extremely important postulate, for it not only provides a conception of entropy which the imagination can grasp, but it provides a connection between thermodynamics and statistical methods.

To formulate the relationship, known entropy and probability relationships are utilized. For two independent systems, A and B, the total entropy is the sum of the two individual entropies S_A and S_B, since entropy is an extensive property of matter. Total probability is the product of constituent probabilities:

$$P = P_A P_B\,.$$

References p. 85

According to Boltzmann's idea:

$$S(P) = S(P_A) + S(P_B)\,.$$

By inspection of these two equations, one might guess that:

$$S(P) = k \ln P\,,$$

where k is some constant whose numerical value must be determined by experiment and is called Boltzmann's constant. This equation is the mathematical expression of Boltzmann's postulate. A proof that this relationship must follow from the two equations for probability and entropy is given in the Appendix.

If, as we are assuming, the probability that applies to a physical system is the maximum value, we make the entropy expression applicable to a physical system by writing:

$$S_T = k \ln P_m\,,$$

where S_T refers to the physically observable entropy of the total ensemble and P_m is the maximum probability.

As we shall see, this relationship makes possible a general method of calculating thermodynamic properties from statistical mechanics. The quantity to be evaluated is P_m, but since logarithm is a monotonic function, the more convenient quantity $(\ln P)_m$ can be calculated instead. In the general procedure, $(\ln P)_m$ is calculated according to whatever statistical method is being used, including as necessary conditions the constancy of total energy and total number of systems. An equation for S is set up and is found to include a very useful quantity called partition function, which is a 'sum-over-states' (or, in the classical treatment, an integral) of a particular function of energy. Partition function is useful because regardless of what probability expression it is developed from, it bears a certain relation to free energy and hence to the other thermodynamic functions.

Let us develop an expression for $(\ln P)_m$ for our Gibbs ensemble. We begin with the expression for $\ln P$ which is: $\ln P = \ln L! - \sum_i \ln L_i!$. The factorial has been approximated by Stirling as $L! = (2\pi L)^{1/2}(L/\mathrm{e})^L (1 + \ldots)$ *, a semiconvergent series with the last term approaching

* A derivation of the first term of the approximation is given in the Appendix (section (*vi*), pp. 83-85).

unity as L is made larger. Taking logarithms and retaining only those terms of the order of L, we obtain $\ln L! = L \ln L - L$. This approximation becomes exact in the limit when L becomes infinite. With the use of this approximation for $L_i!$ also, the expression for $\ln P$ becomes:

$$\ln P = L \ln L - \sum_i L_i \ln L_i \,.$$

For the maximum condition, $\delta \ln P$ must be zero, and so we differentiate the right-hand side of the equation and set it equal to zero. One of the terms which appear is $\sum_i \delta L_i$, which must always equal zero because although the L_i's may vary, their total (L) must remain constant. Upon dropping this term we obtain:

$$\delta \ln P = 0 = -\sum_i \ln L_i \, \delta L_i \,.$$

The solution to this equation must satisfy the following conditions: (a) The quantity $\sum_i \delta L_i$ must always vanish. (b) The quantity $\sum_i u_i \, \delta L_i$ must always vanish. Thus the three equations to be satisfied are:

$$\sum_i \ln L_i \, \delta L_i = 0$$
$$\sum_i \delta L_i = 0$$
$$\sum_i u_i \, \delta L_i = 0 \,.$$

Using Lagrange's method of undetermined multipliers *, we obtain the single equation:

$$\sum_i (\ln L_i + \alpha + \beta \, u_i) \, \delta L_i = 0 \,,$$

where α and β are undetermined parameters. This holds for any variation δL_i, since all the conditions are now included in this one equation. For this reason we set the term in parentheses equal to zero, and we derive the expression:

$$\ln L_i = -\alpha - \beta u_i, \text{ or } L_i = e^{-\alpha} \, e^{-\beta u_i} \,,$$

which applies to the condition of maximum probability. This equation for

* The Lagrange method of undetermined multipliers is a highly useful device for handling maximum or minimum problems in which extra conditions must be fulfilled. Its great usefulness lies in its ability to produce one basic equation containing all the restraints given. The included undetermined parameters, one for each restraint, ensure the attainment of the most general solution to the problem.

References p. 85

L_i enables us to express L and U as:

$$L = \sum_i L_i = \sum_i e^{-\alpha} e^{-\beta u_i} = e^{-\alpha} \sum_i e^{-\beta u_i} ;$$
$$U = \sum_i L_i u_i = \sum_i e^{-\alpha} u_i e^{-\beta u_i} .$$

Because we have three equations in three unknowns (L_i, α, and β), in principle the problem of expressing the L_i's at the condition of maximum probability is completely solved.

We can use the expression for $\ln L_i$ at the condition of maximum probability in our equation for $\ln P$ to arrive at an equation for $\ln P_m$ (which is the same as $[\ln P]_m$):

$$\ln P_m = L \ln L - \sum_i L_i(-\alpha - \beta u_i) = L \ln L + \alpha L + \beta U .$$

Before proceeding with the development of the entropy expression, let us make some observations about the equations for L_i, L, and U, which we have just developed. We can reduce the number of unknowns to two by expressing $e^{-\alpha}$ as its equivalent,

$$L/\sum_i e^{-\beta u_i} .$$

The equations for L_i and U then become:

$$L_i = L \frac{e^{-\beta u_i}}{\sum_i e^{-\beta u_i}} ;$$
$$U = L \frac{\sum_i u_i e^{-\beta u_i}}{\sum_i e^{-\beta u_i}} .$$

Having the equations in this form, we can readily see that the population of any state is a function of its associated energy, and that the total energy is a function of the population of the various states. But it is important to realize that the summations are performed over the states and not just the energy levels. A single energy level (which we can designate as n) is generally associated with more than one state, and so it is not enough to know the energy levels involved – in order to evaluate the sums we need also know the number of states associated with each energy. In the language of quantum mechanics, this number is called the degeneracy of the energy level. In a practical application, each exponential function in a summation must be multiplied by its degeneracy g_n; the summation can then be performed over the energy levels as, for example,

$$\sum_{u_n} g_n \, e^{-\beta u_n} .$$

The influence of the degeneracy accounts for the fact that the lowest energy is not necessarily the most probable in an ensemble, although of course the lowest energy state (the ground state) is more probable than any other single state.

According to Boltzmann's postulate, the entropy of a physical ensemble is represented by:

$$S_T = k \ln P_m .$$

Using our evaluation of $\ln P_m$ we find:

$$S_T = k(L \ln L + \alpha L + \beta U) .$$

Although complete, this expression is extremely difficult to evaluate because the system energies, u_i, are involved in the parameters α and U and must be known exactly.

The quantity β, however, can be evaluated rather easily through thermodynamic considerations. We begin the procedure by substituting for α in the equation for S_T. As we have shown,

$$e^{-\alpha} = L / \sum_i e^{-\beta u_i} ;$$

if for abbreviation we let Σ stand for

$$\sum_i e^{-\beta u_i} ,$$

the expression is written as $e^{-\alpha} = L/\Sigma$, or $\alpha = - \ln L + \ln \Sigma$. Thus the equation for the entropy becomes:

$$S_T = k(L \ln \Sigma + \beta U) .$$

Allowing changes in both β and u_i to affect changes in Σ, we arrive after some mathematical manipulation at an expression for the differential of entropy in terms of a change in internal energy and a change in the energy levels of the individual systems: $dS_T = k\beta(dU - \sum_i L_i \, du_i)$. We know from quantum mechanics that the summation term represents the element of work done by external forces in changing the energy levels, and so we can write: $\mathrm{d}S_T = k\beta(\mathrm{d}U - \delta W)$†. Comparison with the expression that is the combination of the first and second laws of thermodynamics, *i.e.*, $\mathrm{d}U = T\,\mathrm{d}S + \delta W$, identifies β as $1/kT$. Therefore entropy can be expressed as:

$$S_T = L\,k \ln \Sigma + U/T.$$

Using the definition $U/L = \bar{u}$, where $\bar{u}$ is the average energy of one system, we can express the entropy for the single physical system as:

$$S = k \ln \Sigma + \bar{u}/T\,,$$

where, in terms of our evaluation of β, the symbol Σ represents the quantity

$$\sum_i \mathrm{e}^{-u_i/kT}\,.$$

We know that thermodynamic properties depend on internal and external variables; these variables have an influence on the energy level structure and so their effect is accounted for, in these equations, in the quantity Σ.

Free energy expressions can be derived from those for entropy, for we know from thermodynamics that $F = U - TS$. For the entire ensemble:

$$F_T = U - TS_T = -\,LkT \ln \Sigma\,,$$

and for the single system:

$$F = -\,kT \ln \Sigma\,.$$

The quantity Σ is known as the partition function. As mentioned earlier, it may be derived by procedures other than the quantum mechanical method shown here, but its relation to free energy is general.

This simple expression for free energy is very useful, for from free energy all the rest of the thermodynamic properties can be easily calculated. For example, pressure, P, is $-(\partial F/\partial V)_T$; entropy, S, is $-(\partial F/\partial T)_V$; internal energy, U, equals $F + TS$ or, equivalently, $F - T(\partial F/\partial T)_V$; and heat capacity, C, is $(\partial U/\partial T)_V$.

In summary, the important quantity Σ, the partition function, is required for the calculation of free energy and other thermodynamic

---

† See the Appendix for details of the derivation of these equations for the differential of entropy (pp. 78-79).

quantities. And for the evaluation of partition function according to quantum mechanics it is required that the individual energy levels and their degeneracies be known for the system.

The Older (Classical) Approach to Statistical Theory

The expressions for probability, partition function, and thermodynamic quantities that we have derived, have all been based on the ideas of quantum mechanics. These ideas are, of course, relatively modern. It is illuminating to examine the method by which derivations of the same quantities were made before the days of quantum theory: for example, we can see how with the application of classical ideas there was an inconsistency in the calculation of entropy which could not be resolved until the statistical model was re-examined in the light of quantum theory. Let us begin a discussion of the classical method by showing how it is used to express partition function.

The classical method makes use of a very important concept of classical mechanics – the concept of the number of free coordinates or degrees of freedom that a system has. This is simply the minimum number of coordinates necessary to specify completely any position of a mechanical system. These coordinates are usually represented as $q_1, q_2, \ldots q_n$, where n is the number of degrees of freedom. A single particle has three degrees of freedom: x, y, and z; two particles joined by a rigid rod have five: x, y, and z for one, and two angles (latitude and longitude) for the other.

Each free coordinate, q, is associated with a quantity called the conjugate momentum, p, which is defined as a partial derivative of energy with respect to free velocity; the totality of free coordinates and conjugate momenta for a given system completely defines the system at all times. The q's and p's are related through a function called the Hamiltonian, H (which for a conservative system is total energy):

$$\partial q_i/\partial t = \partial H/\partial p_i \; ;$$
$$\partial p_i/\partial t = -\, \partial H/\partial q_i \; .$$

These are the so-called canonical equations of motion *; for a system with n degrees of freedom there are $2n$ such equations. The use of these equations, together with the concept of 'phase space' leads to the development

* These equations are discussed in detail in standard texts on classical mechanics.

References p. 85

of Liouville's Theorem, which is of fundamental importance in the classical statistical method.

The phase space consists of as many Cartesian coordinates as there are free coordinates and conjugate momenta for the system (this number equals $2n$). The system, then, is completely characterized by one point in phase space. For example, for a gas consisting of independent point atoms (helium can be considered such a gas), there is a total of six coordinates and conjugate momenta per atom. If the gas is considered according to the Gibbs concept, one point in a $6N$-dimensional phase space describes the N atoms in a system; the trajectory of the single point describes the state of the system as time passes. For the description of the L systems of N atoms each in a Gibbs ensemble, there are required $6N$ dimensions and L points; the trajectories of these points decribe the state of the entire ensemble with the passage of time. For the description of the gas according to the Maxwell-Boltzmann concept, only six dimensions are required for the phase space, since the number of particles per system is one, but a total of N points is needed to describe a collection of N atoms.

Liouville's Theorem is obtained by consideration of how the point density changes at any given location in the phase space. The difference between the number of points entering and leaving a fixed unit volume within a unit time is the time rate of density change at that place. The position in phase space of each point is defined by its free coordinates and conjugate momenta. These can be considered in pairs, and for simplicity we shall consider a system which is defined by just one pair – a linear harmonic oscillator is such a system. For fixed coordinates the change in point density, ρ, with the passage of time is determined as follows (refer to Fig. 12): consider a rectangle of dimensions dp dq. The

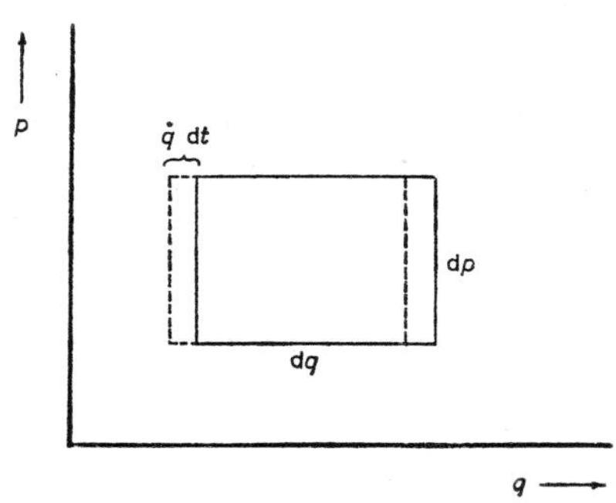

Fig. 12

number of points entering the rectangle per second through the left-hand side, of length dp, is equal to $\rho\,\dot{q}\,\mathrm{d}p$. To a very good approximation the number of points leaving the right-hand side, also of length dp, equals $-\left[\rho\,\dot{q}\,\mathrm{d}p + \partial/\partial q(\rho\dot{q}\,\mathrm{d}p)\right]\mathrm{d}q$, which is the first term of a Taylor expansion. Therefore, the change in the number of points within the rectangle due to those points entering or leaving the sides dp, is $-\partial/\partial q(\rho\dot{q})\,\mathrm{d}p\,\mathrm{d}q$. Similarly, the change in number due to those points passing through the sides dq is $-\partial/\partial p(\rho\dot{p})\,\mathrm{d}p\,\mathrm{d}q$. The sum of these changes, divided by dp dq so as to refer to unit area, is the total change in point density per unit time: $\partial\rho/\partial t = -\partial/\partial q(\rho\dot{q}) - \partial/\partial p(\rho\dot{p})$. This represents the conservation of points in phase space. Performing the product differentiations we obtain:

$$\frac{\partial\rho}{\partial t} + \rho\left[\frac{\partial}{\partial q}(\dot{q}) + \frac{\partial}{\partial p}(\dot{p})\right] + \frac{\partial\rho}{\partial q}\dot{q} + \frac{\partial\rho}{\partial p}\dot{p} = 0\,.$$

By use of the canonical equations the second term may be written:

$$\rho\left[\frac{\partial}{\partial q}\left(\frac{\partial H}{\partial p}\right) - \frac{\partial}{\partial p}\left(\frac{\partial H}{\partial q}\right)\right],$$

which is zero identically. The remaining terms on the left-hand side constitute the total derivative $\mathrm{d}\rho/\mathrm{d}t$, and so the equation is:

$$\mathrm{d}\rho/\mathrm{d}t = 0\,.$$

This result was obtained for a system defined by just one pair of free coordinates and conjugate momenta; if additional pairs of coordinates are required, they must, of course, be included in the calculation. But it can be shown that when all coordinates are considered the total differential $\mathrm{d}\rho/\mathrm{d}t$ always vanishes. Therefore, Liouville's Theorem holds for a phase space of any dimensions.

The equation may be interpreted as saying that as we follow a point in its trajectory, there is no change in the density of its surroundings. In other words, Liouville's Theorem tells us that a point carries, so to speak, a constant density of points with it throughout its trajectory. It appears reasonable, in view of this result, to make the necessary postulate that for equal volumes in the phase space there is equal a priori probability for occupancy. (This postulate is also required in quantum mechanical statistics.) The importance of Liouville's Theorem is that, through this postulate, it indicates that there is a general type of space – phase space –

References p. 85

which is suitable for the calculations connected with all statistical mechanical problems. It should be realized that the theorem is based on: (1) the use of Cartesian coordinates to plot all free coordinates and conjugate momenta, whether originally rectilinear, angular, or otherwise; and (2) the assumption that the points in phase space act independently: the state of a given system does not depend on the state of the other systems around it.

We next find out what dependence the probability (which is proportional to point density) has on energy. To do this we make use, as we did earlier in our discussion of the quantum mechanical approach, of the principle that the actual condition of an ensemble is described by the condition of maximum probability. Using this principle, it is possible to show that probability depends on energy u as the familiar exponential $e^{-u/kT}$. The normalizing factor, called the partition function, $\sum$, is:

$$\sum = \int_{\substack{\text{all } p\\ \text{all } q}} \dots \int e^{-u(p,\, q)/kT}\, dp_1\, dq_1 \dots dp_n\, dq_n\,.$$

The product $q_i p_i$ has the dimension (energy · time), and so $\sum$ has the dimension (energy · time)^n. Here is an example of the shortcomings of the classical theory, for if we try to apply this expression for partition function to thermodynamic calculations, we find ourselves attempting to take the logarithm of a dimensional quantity. An attempt was made at one time to correct this difficulty by introducing a suitable dimensional constant, such as the nth power of Planck's constant h, but the necessity for such an artificial step raises a doubt about the validity of a theory. However, later work in quantum mechanics showed a theoretical basis for the inclusion of h^n. From the Uncertainty Relationship one finds that position and momentum can be simultaneously specified only within a certain limitation given by $\Delta p\, \Delta q = h$. Hence in a phase space of n degrees of freedom, the size of the smallest volume that can be specified is $\Delta p_1\, \Delta q_1 \dots \Delta p_n\, \Delta q_n = h^n$. The final form of the expression for partition function is:

$$\sum = \int_{\substack{\text{all } p\\ \text{all } q}} \dots \int e^{-u(p,q)/kT}\, \frac{dp_1\, dq_1 \dots dp_n\, dq_n}{h^n}\,.$$

This integral expression for partition function, like that derived according to quantum statistics, is of great importance because from it thermodynamic functions can be calculated. The similarity between the two expressions for $\sum$ can be recognized through a consideration of the integration process as follows. Divide all of phase space for a particular collection of systems into thin energy shells. The probability for a particular energy u is then obtained by taking the exponential $e^{-u/kT}$, which represents the density of points, and multiplying it by the shell volume for this energy; (this is permissible since the density is constant throughout the shell). The product is proportional to the total number of points in the shell of energy u, and the sum of the results for all the shells in phase space is the partition function. This integration process is equivalent to the quantum mechanical procedure of summing the exponentials, each of which is multiplied by its degeneracy.

Use of the Classical Approach and the Maxwell-Boltzmann Single Particle System: Calculation of Partition Function and Thermodynamic Properties of a Perfect Gas

In order to make use of statistical theory in calculations of quantities such as thermodynamic functions, it is necessary to evaluate the partition function. We shall first see how this may be done (for a perfect monatomic gas) using the classical approach and the Maxwell-Boltzmann theory, and what limitations this method has.

In the Maxwell-Boltzmann theory, the system is a single particle. If the system is an atom such as helium the phase space is six-dimensional, the coordinates being x, y, and z for position and p_x, p_y, and p_z for momentum. If it is assumed that the N systems in an ensemble are independent, the partition function for the ensemble can be written as the product of the system partition functions:

$$\sum_T = \left[\int_{x \ldots p_z} \cdots \int e^{-u/kT} \frac{dx dy dz dp_x dp_y dp_z}{h^3} \right]^N .$$

We assume for simplicity that there is no potential energy, so that the total energy is kinetic only. This is:

$$K = \frac{m}{2}\left(\dot{x}^2 + \dot{y}^2 + \dot{z}^2\right) = \frac{p_x^2 + p_y^2 + p_z^2}{2m},$$

References p. 85

where, in accordance with the definition of conjugate momentum, $p_x = \partial K/\partial \dot{x} = m\dot{x}$, etc. The total partition function is expressed as:

$$\sum_T = \left[\int_{x \ldots p_z} \cdots \int \mathrm{e}^{-(p_x^2+p_y^2+p_z^2)/2mkT} \frac{\mathrm{d}x \ldots \mathrm{d}p_z}{h^3} \right]^N .$$

Although formidable looking, this expression can be evaluated rather easily. Since the exponent is independent of position, the volume integration can be performed immediately. The exponent is then factored into a product of three terms, each of which is associated with its particular element of momentum:

$$\sum_T = \frac{V^N}{h^{3N}} \left[\int_{-\infty}^{+\infty} \mathrm{e}^{-p_x^2/2mkT} \mathrm{d}p_x \cdot \int_{-\infty}^{+\infty} \mathrm{e}^{-p_y^2/2mkT} \mathrm{d}p_y \cdot \int_{-\infty}^{+\infty} \mathrm{e}^{-p_z^2/2mkT} \mathrm{d}p_z \right]^N .$$

Each integral has the form of the familiar Gaussian integral,

$$\int_{-\infty}^{+\infty} \mathrm{e}^{-\alpha x^2} \mathrm{d}x ,$$

which equals $(\pi/\alpha)^{1/2}$, α in this case being 1/2mkT. The final result for the partition function is:

$$\sum_T = \frac{V^N}{h^{3N}} (2\pi mkT)^{3N/2} .$$

Before proceeding to calculations in which this function is used, we shall discuss the reasonableness of an assumption made in the development of the expression for partition function. We assumed that the actual condition of an ensemble is represented by the most probable configuration, and we accordingly used the condition of maximum probability in our calculation. We can demonstrate informally that this is reasonable by showing that for a special case the most probable configuration is overwhelmingly more probable than any other. Our special case is the density fluctuation of an ideal gas isolated from all external influences.

Let us consider that our ideal gas consists of N atoms in an isolated volume, V, which is divided into compartments, V_i representing the volume of the ith compartment. Since the gas is isolated from all external influences, there is no reason for an atom to prefer one small volume over another, and probability can be expressed in terms of just the three dimensions of ordinary space. The probability, w_i, of finding a chosen atom in V_i is equal to the fractional volume of that compartment:

$$w_i = V_i/V\,.$$

The probability of a configuration in which there are N_1 atoms in V_1, N_2 in V_2, etc., is given by the product of two terms. One term represents the probability of finding given numbers of certain specified atoms in given small volumes, and is written $w_1^{N_1}\, w_2^{N_2} \ldots$ The other term is a permutation factor denoting the number of different ways in which the N atoms can be arranged within the given volume so that *any* N_1 are in V_1, *any* N_2 are in V_2, etc.: this is written $N!/N_1!\, N_2! \ldots$ It should be noted that this factor is written in accordance with the Maxwell-Boltzmann theory which assumes that the atoms are identifiable; reshuffling atoms within a small volume does not change the overall arrangement, but moving an atom from one volume to another does create a different arrangement. For the given configuration, the total number of possible arrangements, called the probability P, is then:

$$P = \frac{N!}{N_1!\, N_2! \ldots}\, w_1^{N_1}\, w_2^{N_2} \ldots.$$

The probability P is already normalized, as can be seen if one realizes that each P is a term in the multinomial expansion of $(w_1 + w_2 + \ldots)^N$, and that the sum of the fractions in parentheses is just one.

By the use of Stirling's approximation for the factorials, the expression for P can be maximized in the familiar manner (as we showed in our discussion of the quantum mechanical method). The only restricting condition is that the total number of atoms be fixed. One finds the expected result that P is a maximum when there is uniform gas density – that is, when the number of atoms, N_i, in a compartment V_i, is exactly proportional to w_i. The expression for maximum probability is:

$$P^0 = \left[\frac{2\pi N}{\Pi_i 2\pi N_i}\right]^{1/2}.$$

Now let us consider that there are fluctuations about this most probable configuration. Suppose the number of atoms in each compartment is changed from its most probable value, N_i^0, by a small amount ν_i, so that $N_i = N_i^0 + \nu_i$. Again the Stirling formula is used in writing the probability expression, but this time the approximation is carried out to include additional terms. As shown in detail in the Appendix, a good approximation for the probability of a configuration with fluctuations ν_i is obtained in the form of a Gaussian function:

$$P = P^0 \, e^{-\sum_i (\nu_i^2/2 \cdot 1/N_i^0)}.$$

It is evident that if the small changes ν_i were zero, P would be P^0. If there were fluctuations, however, P would fall off exponentially as their magnitude increased; this effect is pictured qualitatively in Fig. 13 for the case in which all the N_i^0's and ν_i^2's are the same.

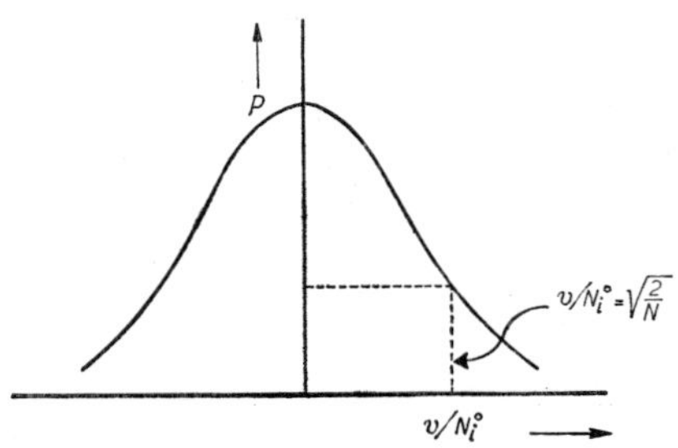

Fig. 13

To acquire an idea of the orders of magnitude involved in the probability formula, let us consider as a special example a unit volume V divided into two equal parts so that $V_1 = V_2 = V/2$ and $N_1^0 = N_2^0 = N/2$, where N is the number of atoms per cubic centimeter. Taking for N its value for an ideal gas at standard conditions, 2.69×10^{19} cm^{-3}, we calculate P^0 to be 2×10^{-10}. Thus it is seen that even the condition of maximum probability is extremely improbable, and at first glance it would appear that any theory which utilizes the most probable configuration to represent the actual condition must be wrong. Confidence in the use of the most probable configuration is restored, however, when the average number of atoms in a cell, in this example $N/2$, is compared with the average magnitude of fluctuations. This latter figure is actually represented by the quantity $\overline{|\nu_i|}$, the average of the absolute values of the various fluctuations, but we shall use instead a sufficiently accurate and more easily calculated quantity, the root mean square,

$$\sqrt{\overline{\nu_i^2}}\,.$$

In our special example there are only two compartments, and so the sum of the fluctuations ν_1 and ν_2 must equal zero; this means that $\nu_1^2 = \nu_2^2$. We also found that $N_1^0 = N_2^0 = N/2$. Putting all this information into the probability formula, we find that the probability for a configuration in which there is a fluctuation ν is:

$$P(\nu) = P^0\,\mathrm{e}^{-2\nu^2/N}\,.$$

The evaluation of $\overline{\nu^2}$ is made by summing, or closely enough by integrating, the product of ν^2 with $P(\nu)$ over the whole range of ν and normalizing. The equation is:

$$\overline{\nu^2} = \frac{\int_{-\infty}^{+\infty} \nu^2 P^0\,\mathrm{e}^{-2\nu^2/N}\,\mathrm{d}\nu}{\int_{-\infty}^{+\infty} P^0\,\mathrm{e}^{-2\nu^2/N}\,\mathrm{d}\nu}\,.$$

The solution is facilitated by the use of two well-known integral evaluations:

$$\int_{-\infty}^{+\infty} \mathrm{e}^{-\alpha x^2}\,\mathrm{d}x = (\pi/\alpha)^{1/2}\,;$$

and

$$\int_{-\infty}^{+\infty} x^2\,\mathrm{e}^{-\alpha x^2}\,\mathrm{d}x = \frac{1}{2}\,\frac{\pi^{1/2}}{\alpha^{3/2}}\,.$$

By identifying α as $2/N$ and x as ν, we find that $\overline{\nu^2} = N/4$. (The value of $\overline{\nu^2}$ is, in general, a function of the most probable number of particles in a compartment – in our special example it is $N_i^0/2$.) The desired quantity, the root mean square, is

$$\sqrt{\overline{\nu^2}} = \sqrt{N}/2\,.$$

Since $N = 2.69 \times 10^{19}$ atoms per cubic centimeter, the average magnitude of fluctuation is about 2.5×10^9, a number which although large is only 5×10^{-11} as large as the average number of atoms in a compartment. In other words, the average fluctuation is immeasurably small and the method of most probable values would be expected to yield the proper results.

In this special example we considered component compartments of large volume. With the same basic expression one may calculate the

References p. 85

fluctuations within tiny volumes taken as parts of a huge body of gas. For instance, the atmosphere can be considered as divided into tiny volumes of magnitude comparable to the cube of the wavelength of visible light. Fluctuations within the tiny volumes would be expected to produce corresponding fluctuations in refractive index. The scattering of light by the atmosphere and, incidentally, the blue color of the sky, can be explained on this basis.

Let us now return to our expression for partition function of a perfect gas, and see how it can be used to calculate thermodynamic quantities. The equation we derived according to classical theory and the Maxwell-Boltzmann system concept is:

$$\sum_T = \frac{V^N}{h^{3N}} \left(2\pi mkT\right)^{3N/2} .$$

We have already seen that free energy is related to partition function as $F = -kT \ln \sum_T$. Hence:

$$F = -NkT\left[\ln V + \ln\left(\frac{2\pi mkT}{h^2}\right)^{3/2}\right],$$

or:

$$F = -NkT\left[\ln V + \frac{3}{2}\ln T + \text{constant}\right].$$

The other thermodynamic functions can be readily calculated from free energy. They are:

$$P = -(\partial F/\partial V)_T = NkT/V ;$$

$$S = -(\partial F/\partial T)_V = \frac{3}{2} Nk(\ln T + 1) + Nk \ln V + \text{constant} ;$$

$$U = F + TS = \frac{3}{2} NkT ;$$

$$C = (\partial U/\partial T)_V = \frac{3}{2} Nk .$$

These results show that the classical theory is successful in derivations of the equation of state for a perfect gas ($PV = NkT$) and the equations for heat capacity and internal energy. The entropy equation, however, yields an incorrect value. At first it seemed that the difficulty must arise from the constant in the expression for free energy, for only in the entropy equation is it retained. Therefore it appeared that the entropy equation could be satisfactorily used to calculate changes in entropy. Experiment showed that this was indeed the case so long as the number of particles was kept constant, but if a change in the number of particles was allowed, a fundamental inconsistency became apparent. This state of affairs became known as the Gibbs paradox.

The Gibbs Paradox: a Basic Deficiency of Classical Statistics

We can describe the Gibbs paradox by means of a simple hypothetical example. Suppose we have a box containing an ideal gas at constant temperature. There is a partition dividing the box in half, and in each half, of volume V, there are N atoms. We know from common experience that if the partition is removed there will be no change in total entropy; a theoretical calculation should predict this result. According to our entropy equation, the volume-dependent part of entropy equals $Nk \ln V$. When the partition is in place (case A) the volume-dependent part of total entropy is the sum of the separate entropies: $S_A = 2Nk \ln V$. When the partition is removed (case B) there is a single volume of size $2V$ containing $2N$ atoms, so that $S_B = 2Nk \ln 2V$ or $2Nk \ln V + 2Nk \ln 2$. In other words, this line of reasoning leads one to expect an increase in entropy equal to $2Nk \ln 2$ when the partition is removed. This quantity is equal to the increase in entropy (the familar entropy of mixing) that we would expect if equal volumes containing equal concentrations of two different gases were mixed. There is, however, no such entropy increase when volumes of the same gas are combined.

It was not until the advent of quantum mechanics that this paradox was successfully resolved and a fundamental error in the classical theory was revealed as its source. Aside from their difference in concept of energy as discrete or continuous, a difference which is mathematically unimportant in considerations of translational energy at ordinary conditions, there is a basic disagreement between classical and quantum mechanics. This difference concerns the distinguishability of particles

References p. 85

and shows up in the expressions for probability and, consequently, for partition function, free energy and entropy.

In the modern method of calculating partition function, the Gibbs concept of systems and ensembles is used in conjunction with quantum mechanics. As it happens, the Gibbs theory is appropriate for application of the correct ideas concerning indistinguishability, but at the time the theory was formulated there was no question raised about distinguishability. Each system was thought of as residing in a macroscopic container that could be 'labelled' if desired. There was no stipulation about a minimum number of particles that had to be present in each container: even just one particle per container would make a perfectly good, completely identifiable system. Indeed, when the containers are thus assumed to hold only one atom each (a very unusual circumstance physically), the calculation of partition function by either classical or quantum mechanical methods yields identical results for room temperature and above. The focal point for the disagreement between quantum and classical mechanics appears when we consider more than one atom per container. Classical theory assumes that the result for N atoms per container is the same as for N single-particle systems; this assumption is considered justifiable by reason of the hypothesis of dynamic independence of the atoms. Quantum theory, although it too assumes dynamic independence, indicates that the N atoms, if placed in one container, do not behave as a superposition of separate systems: this follows from the quantum mechanical concept that an undisturbed particle in a box has the properties of a standing wave, and whereas standing waves in separate boxes can be identified (merely by labelling the boxes), N waves standing together in one box do not have unique identities.

This conceptual difference is evidenced in the classical expression for probability. It contains the term $N!/N_1N_2 \ldots$, a permutation factor denoting the number of different ways in which the N atoms can be arranged. For indistiguishable particles, this factor would be replaced by unity, as it is according to quantum mechanics. But if particles are considered distinguishable, the value of the factor (at moderate and higher temperatures) is very close to just the numerator $N!$, as we shall presently demonstrate. Partition function, which is derived from probability, and free energy, which is derived from partition function, therefore incorporate an incorrect dependence on number of atoms. The specific manner in which this error affects the expression for free energy of a perfect gas can be shown by comparison of the classical expression with the one obtained according to quantum mechanics, and we shall do this later after we develop the quantum mechanical expression. For the moment, it will suffice to say that the overall difference between the two expressions turns out to be proportional to $N \ln N - N$, which is Stirling's approximation for $\ln N!$.

In view of its deficiencies, one may wonder why the classical theory correctly predicts so many of the properties of the ideal gas – even entropy change predicted by the classical method is correct so long as the number of atoms is kept constant. The reason is that although some terms in the classical expression for partition function show the wrong dependency on number of atoms, the number N, though large, is a constant; and because of the nature of the mathematical operations called for, the error does not therefore matter in the calculation of most thermodynamic functions. Details can be readily seen upon closer examination of all the involved equations.

There is one point, though, that can be discussed more fully at this time. We made the statement that the permutation factor $N!/N_1 N_2 \ldots$ is very close to $N!$ at moderate and higher temperatures. The reason the denominator is close to unity can be seen by comparison of the number of available quantum states to the number of atoms for a volume of ideal gas at standard conditions. An easy way to approximate the number of available quantum states makes use of a combination of ideas from the classical and quantum theories. By applying the Uncertainty Principle to the phase space concept, we arrive at the very useful conclusion that within each volume of size h^3 in phase space there is possible only one state for a system. The number of states, $\mathrm{d}Q$, available to a system whose energy lies within some narrow range is thus equal to the volume, expressed in units of h^3, occupied by the corresponding energy shell. This differential volume is six-dimensional, the dimensions being the three position coordinates of ordinary space and three momentum coordinates. Since position does not influence the energy of the ideal gas atom, an integration over the coordinates of physical space can be immediately performed to yield the volume V. What remains is a differential in momentum space. We write:

$$\mathrm{d}Q = \frac{\mathrm{d}p_x \, \mathrm{d}p_y \, \mathrm{d}p_z}{h^3} V .$$

References p. 85

Translational energy depends only on the total momentum p, and so it is convenient to express $\mathrm{d}Q$ in polar coordinates:

$$\mathrm{d}Q = \frac{4\pi\, p^2\, \mathrm{d}p}{h^3} V .$$

Most atoms have an energy that is not far removed from the average, and so for a rough approximation we may say that the total number of available states is equal to the volume enclosed by the energy shell corresponding to the average energy. This is:

$$Q = \frac{\frac{4\pi}{3}\,(p_{av})^3\, V}{h^3} ,$$

where $(p_{av})^3$ is the cube of the momentum that corresponds to the average energy. Average energy is

$$\frac{3}{2}\, kT ,$$

which can be alternately expressed as $p^2/2m$, where m is atomic mass, and so $(p_{av})^3$ is $(3mkT)^{3/2}$. Thus our approximation for the total number of available quantum states is:

$$Q = \frac{\frac{4\pi}{3}\,(3mkT)^{3/2} V}{h^3} .$$

We shall evaluate this quantity by taking the following values: $V = 1\ \mathrm{cm}^3$; $m = 6.7 \times 10^{-24}$ gm (for helium); $h = 6.6 \times 10^{-27}$ erg sec; $kT = 4.1 \times 10^{-14}$ erg(at 300^0 K). This evaluation of Q shows that there are 11×10^{24} available quantum states per cubic centimeter of gas volume. The number of atoms in a cubic centimeter is 2.5×10^{19} at these conditions, so there are available about 10^5 quantum states per atom. This means that two or more atoms will be found in the same quantum state almost never. Since the population of states, N_i, are accordingly almost always either zero or unity, the factorials appearing in the denominator of the permutation factor are almost always unity.

As lower and lower temperatures are approached, the limitations of the classical theory become more significant. Larger numbers of atoms are present in the lower energy states and so, according to the classical concept, the number of possible permutations of the atoms is decreased. In terms of the mathematical expressions, this means that the value of the term $N_1N_2 \ldots$ (which is the denominator of the permutation factor and which we have shown to be very close to unity for moderate and higher temperatures) changes with temperature. Consequently for low temperatures) the classically derived partition function shows an incorrect dependence on not only number of atoms but temperature as well.

The most important conceptual difference between classical and quantum mechanics, then, lies in the issue of particle distinguishability. But also involved in the conceptual difference are the ideas of phase space and energy levels. The classical method for calculating partition function requires, as we have shown, the construction of a complicated phase space in order to 'count' the number of states possessing given energies. According to quantum mechanical ideas, however, the whole notion of phase space is untenable because position and momentum cannot be determined simultaneously; instead, the state of a system is characterized by quantum mechanical parameters. To write a partition function in the quantum mechanical manner, one writes the energy exponential for each level, multiplies each such term by its degeneracy (which is the number of states associated with a particular energy), and adds them all together. The energy level scheme for a particular system is found through the use of Schroedinger's equation.

Thus the new method of statistics applied to the Gibbs ensemble gives in principle a general solution. The chief difficulty is that the technique, even when applied to an ideal gas, is much more involved than is the old classical technique. We shall see how the method may be used by deriving the expression for partition function of an ideal gas.

The Modern Method of Calculating Partition Function for an Ideal Gas

The modern method of calculating the partition function for the ideal gas uses the basic ideas of quantum mechanics as applied to the model of the Gibbs ensemble. Partition function for such an ensemble we found to be the sum over all possible states:

$$\sum_T = \sum_i e^{-u_i/kT} .$$

References p. 85

Each system, containing N atoms, exists in a state, i, of energy u_i. And each of the N atoms in any system exists in some particular atomic state s, the number in each state being designated as N_s; the energy of that state is designated as ε_s. Hence the energy of any system, u_i, is the sum of energies of individual atoms (since ideality is assumed): $\left(\sum_s N_s \varepsilon_s\right)_i$; and the partition function for the ensemble is written:

$$\sum_T = \sum_i \mathrm{e}^{-(\Sigma_s N_s \varepsilon_s)_i/kT} .$$

The summation is performed in the following manner. We select a particular collection of N_s's which, of course, must add up to the constant number of particles, N. The system energies, for this set of N_s's, are obtained from all the possible permutations of these N_s's with the energies ε_s. Now we select a second, different collection of numbers N_s (not merely a permutation of the first) which again add up to N, and calculate a second set of u_i. This process is continued for all possible collections of N_s, and the final sum contains all possible u_i's for all possible N_s's. This summation method is in accordance with the modern assumptions, for the allowed permutations do not include mere exchanges of some atoms in one state with the same number in another; rather, the only permutations counted are those in which there is an exchange of *number* of particles between any two states. Of course, different numbers must be involved in order to create a different state of the system.

If the abbreviation Z_s is used for $\mathrm{e}^{-\varepsilon_s/kT}$, so that $\sum_T$ can be written as $\sum_i (\Pi_s Z_s^{N_s})_i$ or $\sum_i (Z_1^{N_1} Z_2^{N_2} \ldots)_i$, attention is more closely focussed on the individual atomic energy levels. Evaluation is, of course, the same.

Evaluation of $\sum_T$ would not be difficult if it were not for the restraint, $\sum_s N_s = N$. However, there is a way to get around this difficulty. Suppose this restriction were forgotten for the moment; the sum which we have written above as $\sum_i (Z_1^{N_1} Z_2^{N_2} \ldots)_i$ could then be expressed as $\sum_i (Z_1^{N_1})_i \cdot \sum_i (Z_2^{N_2})_i \ldots$, since the value of any N_s has nothing to do with the value of any other. Each sum is of the form $1 + Z_s + Z_s^2 + \ldots$, easily recognized as the expansion for $1/(1 - Z_s)$. The product of all such terms, $\Pi_s \left[1/(1 - Z_s)\right]$, would represent the partition function. Now, starting from this expression, there is a way to include the necessary restriction.

Let

$$\mathrm{f}(\zeta) = \Pi_s \frac{1}{1 - \zeta Z_s},$$

where ζ is merely a parameter. If this function is expanded in terms of ζ through Taylor series and (for convenience) all of the series expansions are multiplied together, there will appear a number of terms in ζ^N whose coefficients are products of Z_s's; the exponents of these coefficients add up to N. If these terms are combined in the form $M\zeta^N$, the M will be precisely the desired partition function*. One does not really have to go through the whole expansion procedure in order to find the value of M: the theory of contour integration in complex variables supplies another method of finding it.

Suppose we have a complex variable, ζ, of the form $\zeta = x + \mathrm{i}y$, where i is the imaginary $\sqrt{-1}$. Let us consider the function c/ζ^n (where n is an integer), which goes to infinity at the point $\zeta = 0$. If an integration is performed around any closed curve (in the complex plane) enclosing the point $\zeta = 0$†, it is found that the integral has a non-zero value only when the integer n is unity; this value is:

$$\oint \frac{c\,\mathrm{d}\zeta}{\zeta} = 2\pi\,\mathrm{i}c\,.$$

In the case of our function,

$$\mathrm{f}(\zeta) = \Pi_s \frac{1}{1 - \zeta Z_s},$$

the quantity we want to extract from the Taylor expansion is the coefficient, M, of the term involving ζ^N. We can make M be the coefficient in the term yielding the non-zero value in a contour integration by dividing our function, $\mathrm{f}(\zeta)$, by ζ^{N+1}: the desired term will then be M/ζ, and the contour integration of the entire function will yield only $2\pi\,\mathrm{i}M$. Thus the partition function, $\sum$, is:

$$\sum = \frac{1}{2\pi\mathrm{i}} \oint \frac{\mathrm{f}(\zeta)}{\zeta^{N+1}}\,\mathrm{d}\zeta\,.$$

This contour integration is difficult to perform; however, a very good approximation is afforded by the so-called method of steepest descent[2]. The basic idea of this method, as applied to our problem, is discussed in

* See the Appendix, p. 81.

† Refer to a text in mathematics for details.

the Appendix; also given, as a simple illustration, is an evaluation of the Gamma function by a technique similar to the method of steepest descent. For the problem at hand the general procedure is as follows. On the real axis the function $f(\zeta)/\zeta^{N+1}$ is real; by ordinary methods of calculus a point, ζ_0, representing a minimum of the function, is found on this axis. A circle, representing the contour, is drawn with its center at the origin and its circumference passing through the point ζ_0 (see Fig. 14). If one

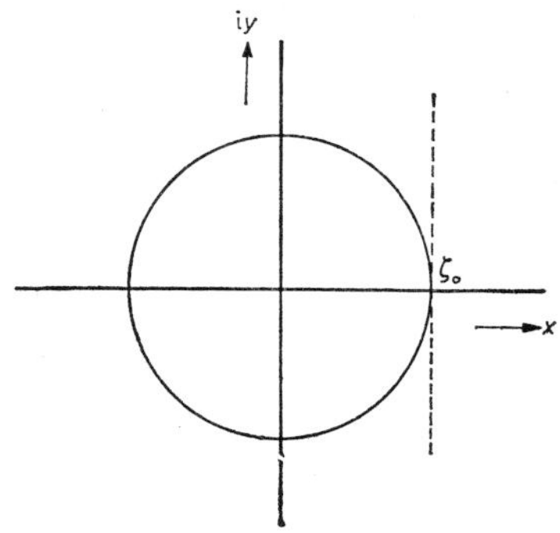

Fig. 14

were to move along the contour through ζ_0, the movement at that point would be in the direction of the imaginary axis, and there, as explained in the Appendix, a *maximum* of the function would be encountered. For large values of N (that is, when the system contains a large number of molecules), this maximum is very sharp, as will be shown; everywhere else on the contour the value of the function is quite small. For this reason, the circular contour can be replaced by a line through ζ_0 parallel to the imaginary axis. A Taylor expansion is made in the direction of the line with only terms up to the second derivative retained since the maximum is so sharp. This new approximate function is integrated over the entire length of the line, yielding the quantity $2\pi\, iM$ or $2\pi i \sum$, where $\sum$ is the desired partition function. Actually, it is necessary to integrate only over a short segment including ζ_0, but because of the nature of the expansion the result is the same and the mathematics are easier if the limits are taken from $-\infty$ to $+\infty$. The mathematical details are as follows:

For convenience write $f(\zeta)/\zeta^{N+1}$ as $e^{g(\zeta)}$. Thus:

$$
\begin{aligned}
g(\zeta) &= \ln f(\zeta) - (N+1)\ln\zeta \\
&= -\sum_s \ln(1-\zeta Z_s) - (N+1)\ln\zeta\,.
\end{aligned}
$$

A minimum in $g(\zeta)$ along the real axis corresponds to a minimum in our original function. Therefore at the minimum:

$$g'(\zeta_0) = \sum_s \frac{Z_s}{1 - \zeta_0 Z_s} - \frac{N+1}{\zeta_0} = 0\,.$$

(The subscript 0 refers to the special value of ζ which produces a minimum on the real axis.) For all practical purposes N and $N + 1$ are the same, and so we may write:

$$N = \sum_s \frac{\zeta_0 Z_s}{1 - \zeta_0 Z_s},$$

which, upon substitution for Z_s is:

$$N = \sum_s \frac{\zeta_0\, e^{-\varepsilon_s/kT}}{1 - \zeta_0\, e^{-\varepsilon_s/kT}}, \quad \text{or} \quad \sum_s \frac{1}{\dfrac{e^{\varepsilon_s/kT}}{\zeta_0} - 1}\,.$$

The summand represents the population of an atomic state. The significance of the parameter ζ can be seen by comparing this expression for population with the one we would obtain using classical treatment. The assumption of identifiability of atoms (the classical idea) leads to the result that the population of an atomic state with associated energy ε_s is proportional to $e^{-\varepsilon_s/kT}$. According to modern theory the system population follows an equivalent law (it is proportional to $e^{-u_i/kT}$), but since only systems and not individual atoms are considered identifiable, the expression for population of an atomic state turns out to be, as we have just shown, the quantity

$$\frac{1}{\dfrac{e^{\varepsilon_s/kT}}{\zeta_0} - 1}\,.$$

In the limit when ζ_0 is very small, this expression becomes proportional to $e^{-\varepsilon_s/kT}$, the classical result.

In the evaluation of ζ_0, the usual assumption will be made that since translational energy levels are so close together, integration can replace summation. Our procedure will be to take the population per atomic

References p. 85

state as given above, multiply by the number of states within a given energy range dε, integrate, and set the result equal to the total number of atoms, N. We have previously shown that the total number of states available to a system whose energy lies within dε is:

$$\mathrm{d}Q = \frac{4\pi p^2 \,\mathrm{d}p}{h^3} V .$$

If we substitute $(2m\,\varepsilon)^{1/2}$ for p and change the variable to the dimensionless ε/kT, we obtain:

$$\mathrm{d}Q = 2\pi \left(\frac{2mkT}{h^2}\right)^{3/2} V \left(\frac{\varepsilon}{kT}\right)^{1/2} \mathrm{d}\left(\frac{\varepsilon}{kT}\right).$$

Collecting all the information and representing ε/kT as x, we express the number N as:

$$N = 2\pi \left(\frac{2mkT}{h^2}\right)^{3/2} V \int_0^\infty \frac{x^{1/2}\,\mathrm{d}x}{\dfrac{\mathrm{e}^x}{\zeta_0} - 1} .$$

It can be seen that when the integral is evaluated, ζ_0 will appear as a function of N/V, the density, rather than of number of atoms alone. For this reason, to be discussed in detail, the Gibbs paradox does not appear in the modern theory.

It is possible to simplify the above equation if ζ_0 can be considered small compared to unity. In this case the integral expression involving x becomes

$$\zeta_0 \int_0^\infty x^{1/2}\,\mathrm{e}^{-x}\,\mathrm{d}x ,$$

a Gamma function with the value $\zeta_0 \dfrac{\sqrt{\pi}}{2}$; thus

$$\zeta_0 = \frac{N}{V}\left(\frac{h^2}{2\pi mkT}\right)^{3/2} .$$

Since ζ_0 is a dimensionless parameter, the quantity in parentheses must have the dimensions of the cube of a length, and we may express ζ_0 as

$$\frac{N}{V}\lambda^3,$$

where λ is a characteristic length equal to $h/(2\pi mkT)^{1/2}$. Using this relationship, let us calculate ζ_0 for an actual gas at ordinary conditions and see if ζ_0 does turn out to be small. For helium at a temperature of 300 ^{0}K and at atmospheric pressure, the numerical values for the needed constants are: $h = 6.6 \times 10^{-27}$ erg sec; $m = 6.7 \times 10^{-24}$ gm; $kT = 4.1 \times 10^{-14}$ erg; $N/V = 2.5 \times 10^{19}$ atoms/cm^3. The quantity λ is calculated to be $\frac{1}{2} \times 10^{-8}$ cm or $\frac{1}{2}$Å. The length of a cube containing, on the average, one atom of the gas is $(2.5 \times 10^{19})^{-1/3}$ cm or 34Å; and so $\zeta_0 = (1/68)^3$, a number very small compared to unity. In order for ζ_0 to be appreciably large – of the order of unity – there would have to be one atom per volume λ^3, which would be one atom per $\frac{1}{2}$Å cube: this would mean that the gas density at this temperature would be greater by a factor of hundreds of thousands. For smaller masses and lower temperatures, λ would be larger, of course. For instance, the characteristic length λ calculated for electrons in a metal at ordinary temperatures would be much larger (though actually, because of the nature of electrons, there is at most one electron per level).

We conclude, then, that at ordinary conditions ζ_0 is small compared to unity and can be approximated as:

$$\zeta_0 = \frac{N}{V}\left(\frac{h^2}{2\pi mkT}\right)^{3/2}.$$

However, it must have a finite value, for its approach to zero is limited by the magnitude of Planck's constant. This evaluation of ζ_0 will be useful in calculating certain functions of ideal gases, for ζ_0 will appear in the quantum mechanical expression for partition function. It will appear in the partition function expression because, as we have shown, $\sum$ is proportional to a contour integral which is to be approximated in terms of ζ_0 by the method of steepest descent.

In order for the method of steepest descent to be satisfactory, the maximum of our function $f(\zeta)/\zeta^{N+1}$ or, equivalently, $g(\zeta)$, along the line

References p. 85

through ζ_0 parallel to the imaginary axis must be very sharp. A measure of the sharpness is given by the second derivative, $g''(\zeta_0)$*, which equals

$$\sum_s \frac{Z_s^2}{(1-\zeta_0 Z_s)^2} + \frac{N+1}{\zeta_0{}^2}.$$

Since N is a very large number and ζ_0 is very small, the second derivative is large and the maximum of the function is sharp. Furthermore, for constant temperature and density, ζ_0 is constant, and so as N increases the sharpness of the maximum becomes more pronounced. Therefore we can say that the method of steepest descent will provide a good approximation of the contour integral.

Now let us develop our expression for partition function. We found that:

$$\Sigma = \frac{1}{2\pi i}\oint \frac{f(\zeta)}{\zeta^{N+1}}\,d\zeta, \text{ or } \frac{1}{2\pi i}\oint e^{g(\zeta)}\,d\zeta\,.$$

In accordance with the method of steepest descent, we substitute for $g(\zeta)$ the Taylor expansion around the point ζ_0, and integrate along the line of steepest descent. The expansion is:

$$g(\zeta) = g(\zeta_0) + g''(\zeta_0)\frac{(\zeta-\zeta_0)^2}{2} + \ldots .$$

(The first derivative at the maximum is zero.) The expression for partition function is thus:

$$\Sigma = \frac{1}{2\pi i}\, e^{\,g(\zeta_0)} \int e^{\left[\frac{(\zeta-\zeta_0)^2}{2} g''(\zeta_0)\right]}\, d(\zeta-\zeta_0)\,.$$

Integration along the line of steepest descent, where in complex terms the variable $(\zeta-\zeta_0)$ is $(x+iy-\zeta_0) = iy$, gives:

$$\Sigma = \frac{1}{2\pi i}\, e^{\,g(\zeta_0)} \int_{-i\infty}^{+i\infty} e^{\,[-y^2/2\cdot g''(\zeta_0)]}\, d(iy) = e^{g(\zeta_0)} \Big/ \sqrt{2\pi g''(\zeta_0)}\,.$$

* Because the differentiation is performed along the real axis, a measure of the sharpness of the *minimum* is obtained from $g''(\zeta_0)$. Due to the nature of complex variables, as discussed in the Appendix, this number also measures the sharpness of the maximum at right angles to it.

For the purpose of calculating thermodynamic functions, it is the logarithm of partition function that is required. From our expression for $\sum$ we see that in good approximation, $\ln \sum = g(\zeta_0)$. By definition:

$$g(\zeta_0) = \ln f(\zeta_0) - (N+1) \ln \zeta_0 \, .$$

From the definition of $f(\zeta)$ we easily calculate that $\ln f(\zeta_0)$ equals $-\sum_s \ln(1 - \zeta_0 Z_s)$ or $-\sum_s \ln(1 - \zeta_0 e^{-x_s})$, where x_s is the abbreviation for ε_s/kT. Thus for our final result we write:

$$\ln \sum_s = -\sum_s \ln(1 - \zeta_0 e^{-x_s}) - (N+1) \ln \zeta_0 \, .$$

Calculation of Thermodynamic Functions by use of Partition Function; Comparison of Results obtained by Classical and Quantum Mechanical Methods

An evaluation of partition function provides a means of calculating thermodynamic properties through its relation to free energy, $F = -kT \ln \sum$. We have just developed an expression for logarithm of partition function by the modern quantum mechanical method; let us now evaluate it for the case of an ideal gas at ordinary conditions, and calculate free energy. The result can be compared with the corresponding evaluation of free energy that we made according to the classical method.

The first term to be evaluated in our expression for $\ln \sum$ is the sum $-\sum_s \ln(1 - \zeta_0 e^{-x_s})$. An excellent approximation for this is the integral of the quantity $-\ln(1 - \zeta_0 e^{-x_s})\, dQ$, where dQ represents the number of states within an energy increment. We previously evaluated dQ as:

$$dQ = 2\pi \left(\frac{2mkT}{h^2}\right)^{3/2} V x^{1/2}\, dx \, .$$

The parameter ζ_0 is, as we have previously shown, very small compared to unity for the case of an ideal gas at ordinary conditions, and so we can approximate $-\ln(1 - \zeta_0 e^{-x_s})$ as $\zeta_0 e^{-x_s}$ *. For the parameter ζ_0 we can use our previously derived evaluation for small ζ_0:

$$\zeta_0 = \frac{N}{V} \left(\frac{h^2}{2\pi mkT}\right)^{3/2} .$$

* This approximation is familiar in the form $\ln(1 - x) = -x$ for very small x.

References p. 85

Substituting and integrating, we obtain for our evaluation of $-\sum_s \ln(1-\zeta_0\, e^{-x_s})$ the number N.

The expression for $\ln \sum$ is so far:

$$\ln \sum = N - (N+1)\ln \zeta_0 ,$$

or:

$$-\ln \sum = -N - N \ln \frac{1}{\zeta_0} .$$

Substituting for ζ_0 we obtain:

$$-\ln \sum = -N - N \ln \left[\left(\frac{2\pi mkT}{h^2}\right)^{3/2} \frac{V}{N}\right] .$$

Then, since $F = -kT \ln \sum$, free energy calculated according to the modern method is:

$$F_{\text{modern}} = -NkT \left\{ 1 + \ln \left[\left(\frac{2\pi mkT}{h^2}\right)^{3/2} \frac{V}{N}\right] \right\} .$$

Let us compare this expression with the one derived according to the classical method:

$$F_{\text{classical}} = -NkT \left\{ \ln V + \ln \left(\frac{2\pi mkT}{h^2}\right)^{3/2} \right\} .$$

For instance, what do these equations predict for the change in free energy with a change in the amount of material considered? According to the modern equation if, say, twice as much material is considered under conditions of constant density (that is, if N and V are both doubled), the total free energy will also be doubled; this prediction is borne out by experiment. The classical equation, however, does not predict this result. In other words, only the modern equation shows the correct relationship between N and V. The mathematical difference between the two equations, $F_{\text{classical}} - F_{\text{modern}}$, is the quantity $kT(N \ln N - N)$ or (since $N \ln N - N$ is Stirling's approximation for $N!$) $kT \ln N!$. (The difference between the corresponding equations for $-\ln \sum$ is $N!$) This is precisely the quantity reflecting the incorrect classical notion of the distinguishability of atoms. Because of the nature of the mathematical operations called for, this difference between the two equations does not matter in the derivation of many thermodynamic functions from free energy. The difference does matter, though, in the derivation of the entropy expression, and this is the reason for the Gibbs paradox, which we discussed in connection with the classical theory.

It may be noted that these equations show that the difference between classical and quantum mechanical theory involves more than just the question of the discreteness of states. In our development of the modern equations for partition function and free energy of an ideal gas at ordinary conditions we made the assumption that the energy levels approach a continuum: we approximated the summation over states by integration. Nevertheless, a significant difference between the classical and modern equations appeared. The discrepancy arises, as we have shown, from the incorrect classical concept of atoms as distinguishable, and so this matter of distinguishability of atoms must be recognized as an important point of difference between classical and quantum mechanical theory.

Another aspect of these equations that should be noted is the appearance of temperature in the form $\ln T$. This presents a difficulty, for when T approaches zero, the logarithm approaches minus infinity: the equations are invalid at low temperatures. The explanation is that these equations, being derived by integration rather than summation, are invalid at low temperatures because then large numbers of atoms exist in a relatively few energy levels. The gas law $PV = NkT$, obtained by an integration process, is likewise invalid at low temperature, even for ideal gases. In such cases the quantum mechanical theory will yield the proper results if summation is used in place of integration, and if the parameter ζ_0, which has an appreciable value in such circumstances, is not approximated under the assumption that it is much smaller than unity.

But even if the exact formulas are used, there is still an apparent difficulty which arises for values of ζ_0 larger than unity: the expression

$$\frac{1}{\dfrac{e^{\varepsilon/kT}}{\zeta_0} - 1},$$

which is the expression for population of an atomic state, may become negative. This situation is strongly reminiscent of the 'negative pressure' predicted by Van der Waals' gas law for gases at low temperatures (see Fig. 15). Just as the gas law paradox is explained through the fact that the material does not remain gaseous throughout its whole pressure-

References p. 85

volume range, but condenses, so that the pressure always remains positive, the paradox concerning the population of atomic states is similarly explained in terms of a 'condensation'. Here, however, the 'condensation is to the ground state of translational energy rather than to a particular state described by the dimensions of ordinary physical space. It may be helpful to think of this 'condensation' in terms of classical ideas: the condensation occurs in momentum space rather than in ordinary space. These points are discussed in greater detail in texts such as Schroedinger [3] and Tolman.[4]

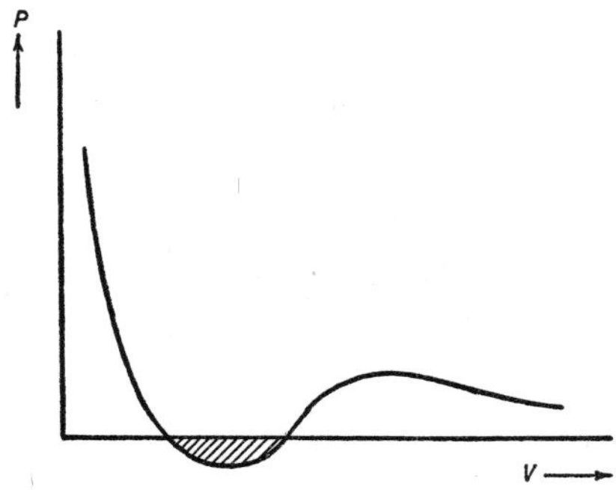

Fig. 15

The population expression we have just discussed appears in the so-called Bose-Einstein statistics, which we have used in our development of an expression for partition function. The assumption is that one particle state can contain many particles. If, instead, the assumption is made that one particle state can hold one particle only, the results are associated with the name Fermi-Dirac statistics. The names are really misnomers, for they give the impression that they represent different statistical methods. Actually, the statistical method is the same for the Fermi-Dirac statistics as for the Bose-Einstein – it is just the stipulation concerning the assignment of particles to particle states that is different.

The Distribution of Dipoles in an Electric Field

An expression for the distribution of gaseous dipoles in an electric field was derived, in the first chapter, by a statistical method whose source was given as a simple analogous problem with known solution. Now that we have discussed the theoretical basis of statistical mechanics, we can derive the distribution expression in a more rigorous manner.

The statistical method used for this problem is the classical Maxwell-Boltzmann method, which is much simpler than the more theoretically sound quantum mechanical statistics and adequate within certain restrictions. Our dipolar material, assumed to be an ideal (non-interacting) gas maintained at an ordinary temperature, meets the requirements. However, our dipolar gas is different in one respect from the perfect gas considered in our discussion of the classical statistical method: its molecules, though non-interacting, are subject to orientation by an applied electric field and so possess potential energy relative to the field. This condition is handled by constructing the phase space to include extra coordinates for the additional degrees of freedom involved. These extra coordinates are those which define the angular position of one atom in the molecule with respect to the other, and the corresponding conjugate momenta.

In our earlier development of the distribution expression, we made the assumption that only the potential energy of the dipoles need be considered in a calculation concerning their orientation. Strictly speaking, however, the energy ε that appears in the theoretical expression for population density, $e^{-\varepsilon/kT}$, is total energy of the molecule, and may be written as $K_1 + K_2 + V$, where K_1 represents the kinetic energy of the center of mass, K_2 is rotational kinetic energy, and V is potential energy.

Partition function is obtained by integration of the expression for population density, $e^{-\varepsilon/kT}$, over the entire phase space. For our problem the phase space is made up of the six coordinates corresponding to translational motion, plus four coordinates corresponding to rotational motion. Translational kinetic energy, being independent of the angular position of the molecule, can have no influence on dipole orientation, and so its integrated value is a constant in so far as dielectric property of the material is concerned. In mathematical terms, we can say that the coordinates of K_1, which are p_x, p_y and p_z, have no angular dependence, and so in the calculation of partition function the integration over all the translation coordinates, (x, y, z, p_x, p_y, p_z), can be immediately performed to yield a constant, C_1. Thus our expression for partition function becomes:

$$\sum = C_1 \int_{\Omega} e^{-(K_2+V)/kT} \, d\Omega \, ,$$

where Ω represents the rotational phase space.

References p. 85

The kinetic energy of rotation of one atom about the center of mass is expressed according to the formula:

$$K_2 = \frac{1}{2} m(\dot{x}^2 + \dot{y}^2 + \dot{z}^2),$$

where m represents the atomic mass and a dot signifies a time derivative. (As will be shown, the result is easily generalized to take into account both atoms.) The distance from an atom to the center of mass is the radius, r, of a spherical area; since this distance is fixed, the atom has only two degrees of freedom, specified by the spherical angles ϑ and φ. The phase space coordinates are, then, ϑ, φ, p_ϑ, and p_φ. By using the transformation $x = r \sin \vartheta \cos \varphi$, $y = r \sin \vartheta \sin \varphi$, $z = r \cos \vartheta$, we can write K_2 as:

$$K_2 = \frac{I}{2} \dot{\vartheta}^2 + \frac{I}{2} \sin^2 \vartheta \dot{\varphi}^2,$$

where I represents the moment of inertia, mr^2. (To take into account both atoms, one simply writes for I the sum $m_1r_1^2 + m_2r_2^2$.) According to the definition of conjugate momentum, $p_\vartheta = \partial K_2/\partial\dot{\vartheta}$ and $p_\varphi = \partial K_2/\partial\dot{\varphi}$; thus we find that $p_\vartheta = I\dot{\vartheta}$ and $p_\varphi = I \sin^2 \vartheta \dot{\varphi}$. The energy K_2 expressed in the proper variables is, then:

$$K_2 = \frac{p_\vartheta{}^2}{2I} + \frac{p_\varphi{}^2}{2I \sin^2 \vartheta}.$$

Partition function is accordingly written:

$$\sum = C_1 \int_\theta \int_\varphi \int_{p_\theta} \int_{p_\varphi} e^{-\frac{1}{2IkT}\left[p_\vartheta{}^2 + \frac{p_\varphi{}^2}{2I\sin^2\vartheta}\right] - \frac{V}{kT}} \mathrm{d}\vartheta \, \mathrm{d}\varphi \, \mathrm{d}p_\vartheta \, \mathrm{d}p_\varphi,$$

or with separation of variables:

$$\sum = C_1 \int_{-\infty}^{+\infty} e^{-\frac{p_\theta{}^2}{2IkT}} \mathrm{d}p_\vartheta \int_\vartheta \int_\varphi \left[\int_{-\infty}^{+\infty} e^{-\frac{p_\varphi{}^2}{2IkT\sin^2\vartheta}} \mathrm{d}p_\varphi \right] e^{-V/kT} \mathrm{d}\vartheta \, \mathrm{d}\varphi .$$

We can evaluate the momentum integrals by using the well-known evaluation:

$$\int_{-\infty}^{+\infty} e^{-x^2/\alpha^2}\,dx = \alpha\sqrt{\pi}\,.$$

If we lump together all constants into a single coefficient, C, our expression for partition function becomes:

$$\sum = C\int_{\vartheta}\int_{\varphi} e^{-V/kT}\sin\vartheta\,d\vartheta\,d\varphi\,,$$

where the term $\sin\vartheta$ results from the integration in the momentum variable p_φ.

The form of this expression has an important consequence, for it enables us to express the partition function as an integral over a spherical surface:

$$\sum = C\int_{s} e^{-V/kT}\,ds\,,$$

where ds is the surface element of the unit sphere, identically $\sin\vartheta\,d\vartheta\,d\varphi$. The form of this expression is precisely that suggested by the analogy of the 'barometric formula'.

(*i*) *A Proof that Entropy is a Logarithmic Function of Probability*

We begin with the expressions (discussed in the text) $P = P_A P_B$ and $S(P) = S(P_A) + S(P_B)$, which in a combined form become $S(P_A P_B) = S(P_A) + S(P_B)$. These equations always hold because of the mathematical nature of probability and the physical nature of entropy. On partial differentiation with respect to P_A we find for the left-hand side of the combined equation:

$$\frac{\partial S(P_A P_B)}{\partial P_A} = \frac{\mathrm{d}S(P_A P_B)}{\mathrm{d}(P_A P_B)} \cdot \frac{\partial (P_A P_B)}{\partial P_A} .$$

Partial differentiation of the right-hand side yields:

$$\frac{\mathrm{d}S(P_A)}{\mathrm{d}P_A} .$$

In summary: $S'(P_A P_B) \cdot P_B = S'(P_A)$, the prime signifying a total derivative. Consideration of the variable P_B leads similarly to the equation: $S'(P_A P_B) \cdot P_A = S'(P_B)$. Comparison of the two equations shows that $P_B S'(P_B) = P_A S'(P_A) = \text{constant}$. It follows that $S(P)$, or S, is $k \ln P$, where k is to be determined by experiment.

(*ii*) *Derivation of an Expression for the Differential of Entropy*

We have expressed the entropy of a Gibbs ensemble as $S_T = k(L \ln \sum + \beta U)$, where the partition function $\sum$ represents the summation $\sum_i \mathrm{e}^{-\beta u_i}$, u_i is the energy level of an individual system, U is total internal energy, and β is a parameter as yet undetermined. We shall assume initially that the energy levels are variable in a purely mathematical sense. A change in the entropy, $\mathrm{d}S_T$, can be written:

$$\mathrm{d}S_T = k \left[L \frac{\mathrm{d}\sum}{\sum} + U\,\mathrm{d}\beta + \beta \mathrm{d}U \right] .$$

From the definition of $\sum$ it follows that

$$L \frac{\mathrm{d}\sum}{\sum} = L \frac{-\sum_i \mathrm{e}^{-\beta u_i} (\beta \, \mathrm{d}u_i + u_i \, \mathrm{d}\beta)}{\sum},$$

which is also

$$-L\beta \frac{\sum_i \mathrm{e}^{-\beta u_i} \, \mathrm{d}u_i}{\sum_i \mathrm{e}^{-\beta u_i}} - L \, \mathrm{d}\beta \frac{\sum_i \mathrm{e}^{-\beta u_i} \, u_i}{\sum_i \mathrm{e}^{-\beta u_i}}.$$

The first term may be written

$$-\beta \frac{\sum_i L \, \mathrm{e}^{-\beta u_i} \, \mathrm{d}u_i}{\sum_i \mathrm{e}^{-\beta u_i}},$$

which is recognized as $-\beta \sum_i L_i \, \mathrm{d}u_i$ by the definition of L_i. Through the definition of U the second term may be written $-U \, \mathrm{d}\beta$. After collecting all this information in the form $\mathrm{d}S = k\beta \left[\mathrm{d}U - \sum_i L_i \, \mathrm{d}u_i\right]$, we are ready for its physical evaluation. The second term represents the results of changing the external contraints: the energy levels for an ideal gas depend on, for example, the volume of the container. When each energy level shift is multiplied by the population of that level and the sum is taken, the result represents the element of work, δW, done on the gas. Thus the entropy change appears in the form $\mathrm{d}S_T = k\beta(\mathrm{d}U - \delta W)$.

(*iii*) *Fluctuations around the Most Probable Spatial Distribution of Ideal Gas in a Container*

The starting point is the expression for probability (in logarithmic form) for a distribution of N molecules with N_1 being in the first compartment of fractional volume w_1, N_2 in the second of fractional volume w_2, etc.:

$$\ln P = N \ln N - N + \tfrac{1}{2} \ln (2\pi N) - \sum_i \left[N_i \ln N_i - N_i + \tfrac{1}{2} \ln (2\pi N_i)\right] + \sum_i N_i \ln w_i \, .$$

(This equation is derived from the general probability equation through the use of Stirling's approximation, $A! = (2\pi A)^{1/2} (A/\mathrm{e})^A$. When the logarithm of the resulting probability expression is written, the results are carried through to second approximation because, as will be seen, first approximation terms all cancel when fluctuations are considered.)

References p. 85

Now let us consider a distribution such that the number of molecules in each compartment, N_i, is very close (in a relative sense) to its most probable value N_i^0; that is, $N_i = N_i^0 + \nu_i$, where ν_i is very small compared to N_i^0. In the expression for ln P there are two summations involving N_i. Let the first of these be designated as $\mathrm{f}(N_i)$. In terms of most probable values N_i^0 and variations ν_i, $\mathrm{f}(N_i)$ is in Taylor expansion:

$$\mathrm{f}(N_i) = \mathrm{f}(N_i^0) + \mathrm{f}'(N_i^0)\,\nu_i + \mathrm{f}''(N_i^0)\,\frac{\nu_i^2}{2}\,.$$

Since:

$$\mathrm{f}(N_i^0) = N_i^0 \ln N_i^0 - N_i^0 + \tfrac{1}{2}\ln(2\pi N_i^0)\,,$$

then:

$$\mathrm{f}'(N_i^0) = \ln N_i^0 + \frac{1}{2N_i^0}\,,$$

which for all practical purposes is ln N_i^0 alone; and

$$\mathrm{f}''(N_i^0) = \frac{1}{N_i^0}\,.$$

This second derivative is of much greater importance than the first derivative because in the Taylor expansion $\mathrm{f}'(N_i^0)$ is multiplied by ν_i whereas $\mathrm{f}''(N_i^0)$ is multiplied by ν_i^2. (Although very small compared to N_i^0, each ν_i may be much larger than unity.)

Collecting all this information together we find so far:

$$\ln P = N\ln N - N + \tfrac{1}{2}\ln(2\pi N) - \sum_i \left[N_i^0 \ln N_i^0 - N_i^0 + \tfrac{1}{2}\ln(2\pi N_i^0)\right]$$
$$+ \sum_i N_i^0 \ln w_i + \sum_i \nu_i \ln w_i - \sum_i \nu_i \ln N_i^0 - \sum_i \frac{\nu_i^2}{2}\cdot\frac{1}{N_i^0}\,.$$

Since the condition of maximum probability is the condition of uniform density, N_i^0 equals Nw_i, and we can write:

$$-\sum_i \nu_i \ln N_i^0 = -\sum_i \nu_i \ln w_i - \sum_i \nu_i \ln N\,.$$

The last term is zero, as can be seen when it is written in the form $-\ln N \sum_i \nu_i$; and the remainder cancels a term in the equation for ln P. The final form of the equation is:

$$\ln P = \ln P^0 - \sum_i \frac{\nu_i^2}{2}\cdot\frac{1}{N_i^0} \text{ or } P = P^0\,\mathrm{e}^{-\Sigma_i \frac{\nu_i^2}{2}\cdot\frac{1}{N_i^0}}\,.$$

(*iv*) *A Demonstration by Simple Example that the Coefficient of ζ^N is the Partition Function*

Consider only three atomic energy levels and two atoms. Then the function

$$\prod_s \frac{1}{1 - \zeta Z_s}$$

is

$$\frac{1}{1 - \zeta Z_1} \cdot \frac{1}{1 - \zeta Z_2} \cdot \frac{1}{1 - \zeta Z_3},$$

or in expanded form,

$$(1 + \zeta Z_1 + \zeta^2 Z_1{}^2 + \ldots)(1 + \zeta Z_2 + \zeta^2 Z_2{}^2 + \ldots)(1 + \zeta Z_3 + \zeta^2 Z_3{}^2 + \ldots).$$

The total coefficient of of ζ^2 is $Z_1{}^2 + Z_2{}^2 + Z_3{}^2 + Z_1 Z_2 + Z_1 Z_3 + Z_2 Z_3$. This result for partition function is in accordance with quantum mechanical theory: the only macroscopic states considered are those in which there are two atoms in any one level, or one atom in one level and the other in another level. Also, the mixed terms do not contain the coefficient two, which would imply that simply exchanging atoms creates a new state.

(*v*) *The Method of Steepest Descent as Used in the Calculation of Partition Function*

Suppose ζ is a complex variable, written $\zeta = x + iy$, i being the imaginary quantity $\sqrt{-1}$. A function of this complex variable, $A(\zeta)$,may be written $A(\zeta) = u(x, y) + iv(x, y)$, where both u and v are real quantities. According to the theory of complex variables, if this function is differentiable with respect to ζ (the requirement for differentiability is that the limit of

$$\frac{A(\zeta + \Delta\zeta) - A(\zeta)}{\Delta\zeta}$$

be the same regardless of the direction of $\Delta\zeta$ in the complex plane), then

References p. 85

the following equations will be satisfied:

$$\text{A. } \partial \mathrm{u}/\partial x = \partial \mathrm{v}/\partial y \ ;$$

$$\text{B. } \partial \mathrm{u}/\partial y = - \partial \mathrm{v}/\partial x \ .$$

These are known as the Cauchy-Riemann differential equations. By differentiating equation A with respect to x and B with respect to y, one finds on comparison that:

$$\partial^2 \mathrm{u}/\partial x^2 + \partial^2 \mathrm{u}/\partial y^2 = 0 \ ;$$

and similarly it is found that:

$$\partial^2 \mathrm{v}/\partial x^2 + \partial^2 \mathrm{v}/\partial y^2 = 0 \ .$$

(These equations show that the functions u and v are both 'harmonic'.)

Since it can be shown that all polynomials and rational fractions are differentiable, the Cauchy-Riemann equations can be applied to the function $\mathrm{f}(\zeta)/\zeta^{N+1}$, which is of special interest to us because it is considered in the discussion of partition function in the text. We shall call this function $\mathrm{h}(\zeta)$ for convenience.

By definition, $\mathrm{f}(\zeta)$ is:

$$\prod_s \frac{1}{1 - \zeta Z_s} \ ;$$

it can be seen by inspection that on the real axis $\mathrm{h}(\zeta)$ is real and has the value u, the quantity v and all its derivatives with respect to x being zero. Suppose that on the real axis there is found a point, ζ_0, where $\mathrm{h}(\zeta)$, or u, is a minimum in the x direction. This means that the second partial derivative $\partial^2 \mathrm{u}/\partial x^2$ must be positive at ζ_0. Then, because $\partial^2 \mathrm{u}/\partial x^2 = - \partial^2 \mathrm{u}/\partial y^2$, the sign of the second derivative along a line through ζ_0 parallel to the y axis must be negative. In addition, since $\partial \mathrm{v}/\partial x = 0$, we know from the Cauchy-Riemann differential equations that $\partial \mathrm{u}/\partial y = 0$. We conclude that since $\partial \mathrm{u}/\partial y$ is zero and $\partial^2 \mathrm{u}/\partial y^2$ is negative, it must be that u is at a *maximum* in the y direction at the point ζ_0. It is, of course, a minimum in the x direction: because of their harmonic character, the functions u and v never possess absolute proper maxima.

Since u is a minimum in the x direction at ζ_0, we know that $\partial \mathrm{u}/\partial x = 0$ at ζ_0; therefore, by reason of the Cauchy-Riemann relations, the first derivative of v in the y direction, $\partial \mathrm{v}/\partial y$, must also equal zero at that point.

The second derivative of v in the y direction vanishes everywhere along the real axis: we know this because $\partial^2 v/\partial x^2 + \partial^2 v/\partial y^2 = 0$ and $\partial^2 v/\partial x^2$ vanishes along the real axis. Thus we can say that not only v but its first and second derivatives with respect to y vanish at ζ_0. This means that in the vicinity of ζ_0 the absolute value of the function h(ζ), even to second approximation, is merely u.

The situation can be pictured in this way: suppose we construct a circle with its center at the origin of complex coordinates and with its circumference passing through ζ_0. If we were to follow along this contour, we would be moving in the y direction at ζ_0 and would there encounter a maximum in the functions u and h(ζ). As is well discussed in Schroedinger's book,[5] a more complete consideration of h(ζ) along the circle shows that the function is not only a maximum at ζ_0 but goes to zero rapidly everywhere else; furthermore as, we have shown, the maximum becomes sharper as N, the number of particles, increases.

Clearly, then, the great contribution to the circuit integral comes from that short segment of the circle that includes the point ζ_0. For all practical purposes the integration may be performed along a straight line tangent to the circle at ζ_0. This is the basis for the method of steepest descent.

(*vi*) *An Evaluation of the Gamma Function by a Technique similar to the Method of Steepest Descent*

The method of steepest descent, used in evaluating partition function, is based on the idea that a circuit integral is approximated by an evaluation of that short segment of the closed curve which includes the point at which the function is a maximum. The segment is evaluated as the integral of a Taylor expansion of the function around its maximum value. This method (whose application is in terms of complex variables) is illustrated by a similar method used to evaluate the Gamma function, or factorial of a number. A general integral expression for the factorial is approximated as the integral of a Taylor expansion (to second derivatives) of the function around its maximum value. The result obtained is found to be identical to the first term of Stirling's approximation for the factorial.

The general integral which represents the factorial of a number, n, is

$$\int_0^\infty x^n \, e^{-x} \, dx \, .$$

References p. 85

The validity of this representation may be seen through these considerations: (a)

$$\int_0^\infty x^{n+1}\, \mathrm{e}^{-x}\, \mathrm{d}x = (n+1) \int_0^\infty x^n\, \mathrm{e}^{-x}\, \mathrm{d}x\,,$$

as required by the fact that $(n + 1)! = (n + 1)\, n!$; and (b) when $n = 1$, the integral is evaluated as unity, as required.

Our method provides a means to approximate this integral. The first step is to write $x^n\, \mathrm{e}^{-x}$ as one term, $\mathrm{e}^{\mathrm{g}(x)}$ where $\mathrm{g}(x)$ stands for $-x + n \ln x$. In this notation the integral to be evaluated is

$$\int_0^\infty \mathrm{e}^{\mathrm{g}(x)}\, \mathrm{d}x\,.$$

If the main contribution to the integral comes from the vicinity of the maximum in $\mathrm{g}(x)$, we can approximate the integral by evaluating $\mathrm{g}(x)$ in the exponent as the Taylor expansion of this function around its maximum value, $\mathrm{g}(x_0)$, and integrating from $-\infty$ to $+\infty$. There is only one maximum of $\mathrm{g}(x)$, and this occurs where $\mathrm{g}'(x)$ is zero; thus we write $\mathrm{g}'(x_0) = -1 + n/x_0 = 0$, from which it is seen that $x_0 = n$. The second derivative, $\mathrm{g}''(x_0)$, equals $-n/x_0^2$ or $-1/n$. The accuracy of the approximation depends on the sharpness of the function's maximum. In the present example, the second derivative, which is $-1/n$, indicates that the maximum unfortunately becomes more diffuse as n increases; this is not the case in the partition function problem. The form $1/n$ is advantageous, however, in that higher derivatives of $\mathrm{g}(x)$ contain higher powers of $1/n$, so that the Taylor expansion in the vicinity of the maximum need not include terms containing derivatives higher than the second.

To summarize, the quantities pertinent to the Taylor expansion are:

$$\mathrm{g}(x) = -x + n \ln x$$

$$\mathrm{g}'(x_0) = 0$$

$$x_0 = n$$

$$\mathrm{g}''(x_0) = -1/n$$

$$\mathrm{g}(x_0) = -n + n \ln n$$

Thus:

$$g(x) = -n + n \ln n - \frac{(x - x_0)^2}{2} \cdot \frac{1}{n}$$

Our evaluation of $n!$ is:

$$n! = \int_0^\infty e^{g(x)}\, dx$$

$$\cong e^{n \ln n - n} \int_{-\infty}^{+\infty} e^{-(x - x_0)^2/2n}\, dx = (2\pi n)^{1/2}\, (n/e)^n\, .$$

This is the first term of Stirling's approximation for the factorial.

REFERENCES

1. SCHROEDINGER, E., *Statistical Thermodynamics*, Cambridge Univ. Press, 1946.
2. DEBYE, P., *Math. Ann.*, 67 (1909) 535.
3. SCHROEDINGER, *loc. cit.*
4. TOLMAN, R. C., *The Principles of Statistical Mechanics*, The Clarendon Press, Oxford, 1938.
5. SCHROEDINGER, *loc. cit.*

CHAPTER 3

The Dielectric Properties of Condensed Phases

Information about the structure of molecules can be obtained from measurements of dielectric properties of a material, as we have demonstrated in Chapter 1. However, the initial theory relating polarization of a molecule to dielectric constant, which we discussed, was developed for gases at low volume concentrations: the theory is valid only when there is no appreciable effect due to interaction between molecules and when the field acting on each molecule can be assumed to be the macroscopic applied field. More highly condensed phases, in which the individual molecules are subjected to both chemical and electrical forces of interaction, cannot be handled with this theory. For this reason investigators soon began to attempt to develop a theory of dielectric property which would be applicable to condensed phases – a theory which could perhaps be utilized to shed light on the workings of these inner forces responsible for the tremendously varied properties of condensed phases.

The Clausius-Mossotti, or Lorentz-Lorenz Equation

Four investigators, working separately, made the first successful attempts to improve the theoretical equation. Their aim was to make an estimate of the actual field acting on an individual molecule, for it is this field that induces polarization of the molecule. The four men derived an equation with which their names are now associated: it is called the Clausius-Mossotti, or the Lorentz-Lorenz equation [1]. This equation takes into account the electrical forces arising from induced dipoles, although it neglects chemical forces existing between molecules, as well as long-range electrical forces due to permanent dipoles.

Let us begin the development of this equation by estimating the field acting on a molecule. We consider a particular molecule of a condensed medium which is under the influence of a homogeneous applied field, $\tilde{E}$. Let us circumscribe, around our molecule, a hypothetical sphere of radius r: the sphere is to be sufficiently large so that the material outside it may be treated as homogeneous. We assume that the medium is isotropic – *i.e.*, that the resulting polarization, $\tilde{P}$, is parallel to the field $\tilde{E}$. The actual field, $\tilde{F}$, acting on our molecule, will be made up of three superimposed fields:

$$\tilde{F} = \tilde{E} + \tilde{F}_0 + \tilde{F}_i \,,$$

where $\tilde{F}_0$ represents the field due to the dipoles in the entire volume outside of the chosen spherical region, and $\tilde{F}_i$ is the field due to the dipoles within the sphere surrounding the given molecule.

The field $\tilde{F}_0$ must arise from effective charge residing on the spherical surface, since everywhere else beyond the interface the dipolar charges counterbalance each other. We know that the effective surface charge density, σ, is equal to the normal component of the volume polarization. Thus $\sigma = P_n = P \cos \vartheta$, where ϑ is the polar angle measured from the $\tilde{E}$ direction. The charge on an element of area ds, is thus $P \cos \vartheta \, \mathrm{d}s$, and the resulting field at the center of the sphere is $P \cos \vartheta \, \mathrm{d}s/r^2$. We know, by reasons of symmetry, that only the component in the $\tilde{E}$ direction is non-vanishing; this component is

$$\frac{P \cos \vartheta \, \mathrm{d}s}{r^2} \cos \vartheta \,.$$

Since $P \cos \vartheta$ is a function of only the polar angle ϑ, we may take for the element of area, ds, the expression $2\pi r^2 \sin \vartheta \, \mathrm{d}\vartheta$. Finally, upon integrating over the sphere, we obtain:

$$F_0 = \int_0^\pi \frac{P \cos^2 \vartheta}{r^2} 2\pi r^2 \sin \vartheta \, \mathrm{d}\vartheta = \frac{4\pi}{3} P \,.$$

The field $\tilde{F}_i$, which arises from the dipoles within the sphere of material, can be calculated through potential energy considerations of our central (reference) dipole. The procedure is as follows. Let us suppose that we have some charge distribution located at a point described by the coordinates x_p, y_p, z_p of a cartesian coordinate system; then there will be a resultant potential Φ at any given point. If in this coordinate system our reference dipole is located with its negative end at the origin and its positive end at the point 0, 0, h (as shown in Fig. 16), then the potential energy of the dipole is equal to $-e\Phi(0, 0, 0) + e\Phi(0,0, h)$. In good approximation, this is $eh(\partial\Phi/\partial z)_{origin}$, which is $m(\partial\Phi/\partial z)_{origin}$, where m is dipole moment. Now if we suppose that the charge distribution at

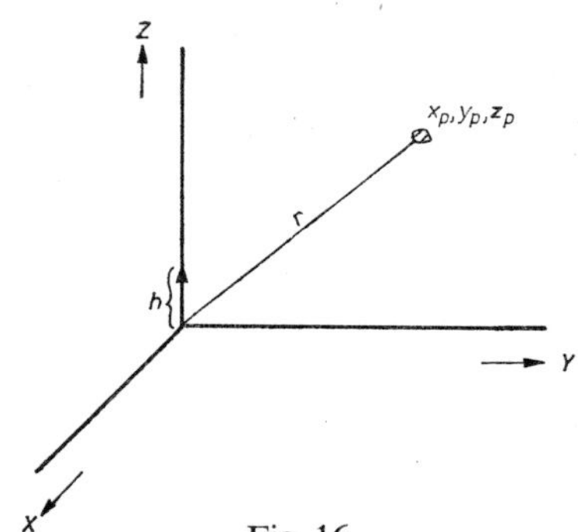

Fig. 16

x_p, y_p, z_p is an identical dipole, oriented in the same direction, we know that the potential which it produces at any point x, y, z can be written as:

$$\Phi = \frac{m\,(z - z_p)}{\left[(x - x_p)^2 + (y - y_p)^2 + (z - z_p)^2\right]^{3/2}}\,.$$

In particular if we perform the indicated differentiation of Φ at the origin, then on substituting we find that the potential energy of the reference dipole at the origin due to the dipole at x_p, y_p, z_p is:

$$m\left(\frac{\partial\Phi}{\partial z}\right)_{\text{origin}} = \frac{m^2}{r^3}\left[1 - \frac{3z^2}{r^2}\right].$$

Now, the factor $(\partial\Phi/\partial z)_{origin}$ is simply the z component of the gradient of the potential and thus represents the field strength $-E_z$ at the origin. Hence in calculating the potential energy of the dipole, we are calculating the component of the field in the dipole direction (multiplied by m). Moreover, if we make the assumption that the dipole molecules within our sphere are arranged in cubic symmetry, we see that at the central dipole there can be no resultant field strength in the x or y directions, and that the field in the dipole direction constitutes the entire field $\tilde{F}_i$. Thus if we can calculate the total potential energy of the central dipole due to all the surrounding dipoles, we shall have calculated a quantity proportional to the field strength $\tilde{F}_i$.

To calculate total potential energy of the central molecule due to the surrounding symmetrically-arranged dipoles, we write the coordinates in terms of a 'lattice distance' a. Thus the coordinates of a dipole are $ja = x_p$, $ka = y_p$, and $la = z_p$, where j, k, and l are integers. The total potential energy, V, of the central dipole in terms of these coordinates is:

$$V = \frac{m^2}{a^3} \sum_{j,k,l} \frac{j^2 + k^2 + l^2 - 3l^2}{(j^2 + k^2 + l^2)^{5/2}}\,.$$

Because of the cubic symmetry, there will be a dipole located at each point whose coordinates are permutations of ja, ka, and la. Therefore, for a given combination of integers, the denominators of the three summed terms will all be the same, but the sum of the numerators will be $(j^2 + k^2 + l^2 - 3l^2) + (k^2 + l^2 + j^2 - 3j^2) + (l^2 + j^2 + k^2 - 3k^2)$, which equals zero identically. All the terms for the sphere may be grouped in this manner into a series of identically vanishing expressions, yielding the final result that for the case of cubic symmetry, the field $\tilde{F}_i$ vanishes.

The magnitude of the true field acting to polarize the molecule is therefore calculated to be:

$$F = \frac{4\pi}{3} P + E\,.$$

The volume polarization, P, is related to the microscopic field strength acting on the molecule according to the equation $P = n\alpha F$, where n is the number of dipoles per unit volume and α is the polarizability per molecule. Substituting for F the above estimation, we find:

$$P = n\alpha \left[\frac{4\pi}{3} P + E\right], \text{ or } P = \frac{n\alpha}{1 - \frac{4\pi}{3} n\alpha} E\,,$$

This evaluation of P enables us to derive an expression for dielectric constant through the use of the general relationships between displacement D, dielectric constant ε, and polarization P. We know that $D = E + 4\pi P$. Using our evaluation of P, we find:

$$D = E\left[1 + \frac{4\pi n\alpha}{1 - \frac{4\pi}{3} n\alpha}\right].$$

References p. 149

But since $D = \varepsilon E$, the bracketed factor represents ε. This can be written in the form:

$$\varepsilon = \frac{1 + \frac{2}{3}(4\pi n\alpha)}{1 - \frac{1}{3}(4\pi n\alpha)}$$

or, alternately:

$$\frac{\varepsilon - 1}{\varepsilon + 2} = \frac{4\pi}{3} n\alpha ,$$

which is the Clausius-Mossotti or Lorentz-Lorenz equation. It may be noted that in the case of a gas, this equation, for all practical purposes, reduces to the earlier expression, $\varepsilon - 1 = 4\pi n\alpha$.

In the development of this equation it was assumed that the dipoles are arranged in cubic symmetry. The equation becomes applicable to an isotropic liquid if the symmetry is assumed to be spherical, and it can be shown that the same equation is obtained with this assumption. The condition of spherical symmetry can be described in this way: if one dipole is located at a point whose x, y, and z coordinates with respect to the central dipole are a, b, and c, then other dipoles are located at those positions whose coordinates are permutations of a, b, and c (including both positive and negative signs). The coordinates a, b, and c may have any value (for the case of a liquid) and generally are not whole multiples of some basic unit of length, like the coordinates ja, ka, and la pertaining to cubic symmetry. Nevertheless, the coordinates of all the dipoles can be arranged so as to yield a vanishing internal field, F_i, and so the final equation is the same as the one derived with the assumption of cubic symmetry. Because of the assumption of spherical symmetry, the total field strength, F, must be considered as uniform throughout the liquid (for constant applied field E) and so should more properly be called an 'average' field.

When measurements are made at optical frequencies where the permanent dipoles do not turn with the field and consequently exert no effect, the Clausius-Mossotti equation may be expressed in terms of refractive index, r, rather than dielectric constant. The Maxwell relation $\varepsilon = r^2$ is utilized, and the equation becomes:

$$\frac{r^2 - 1}{r^2 + 2} = \frac{4\pi}{3} n\alpha(\nu) ,$$

in which polarizability is a function of frequency ν. This equation is applicable to condensed media as well as to gases (for example, it holds very well for benzene in either the vapor or liquid state at the frequency of yellow light), whereas the older formula,

$$\frac{r^2 - 1}{3} = \frac{4}{3} n\alpha(\nu) ,$$

is adequate to describe only the gaseous state. The reason for this is that in the gaseous state dielectric constant is so close to unity that $\varepsilon + 2$ is practically 3.

The Clausius-Mossotti equation proves to be inadequate in some applications. For example, its predictions concerning permanent polarization can be quite erroneous. According to the equation, dielectric constant must approach infinity as

$$\frac{4\pi}{3} n\alpha$$

approaches unity; but since $(\varepsilon - 1)E = 4\pi P$, it is apparent that at values of ε approaching infinity, there would be a finite polarization even with an infinitesimal applied field E. In other words, the equation predicts the phenomenon of permanent polarization. It is possible to calculate for a given material the 'critical temperature', at which the term

$$\frac{4\pi}{3} n\alpha$$

would just equal unity, and below which permanent polarization would consequently be expected. For the purpose of making such a calculation, we write α as the sum of deformation and orientation polarizabilities, obtaining the equation in the form:

$$\frac{\varepsilon - 1}{\varepsilon + 2} = \frac{4\pi}{3} n \left(\alpha_D + \frac{\mu^2}{3kT} \right) ,$$

where α_D is the distortion or deformation polarizability and the term $\mu^2/3kT$ represents the orientation polarizability. Both α_D and μ can be

References p. 149

obtained from existing data on gas phase measurements of the dependence of α on temperature, as discussed in Chapter 1. Using such information in the equation, we find for liquid water a critical temperature of over a thousand degrees Kelvin, a result which erroneously implies that water is permanently polarized at ordinary temperatures.

A calculation of this critical temperature can be made by a more general procedure. When a material exhibits polarization due to both induced and oriented dipoles, the graph $P = \mathrm{f}(F)$ (see Fig. 17) is the result of the two effects: to the linear distortion polarization, $n\alpha_D F$, is added the orientation polarization, $n\alpha_0 F$. But the true field, F, depends in turn on P, as expressed in the equation $F = E + \beta P$, where, according

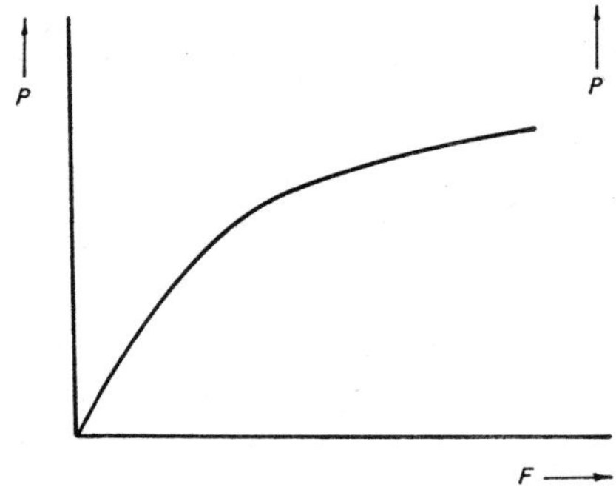

Fig. 17

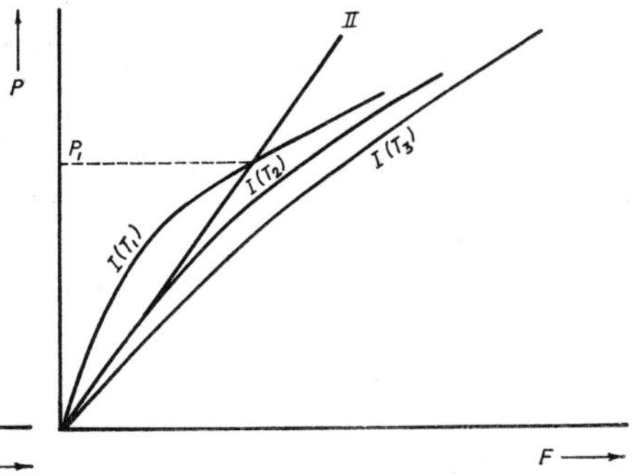

Fig. 18

to the Clausius-Mossotti idea, β is a constant related to the system. In the absence of an external field, $F = \beta P$, or, equivalently, $P = (1/\beta)F$, which is the equation of a straight line. Thus we have two equations for P:

$$\text{I. } P = \mathrm{f}(F)\ ;$$
$$\text{II. } P = (1/\beta)F \text{ when } E = 0\ .$$

Solutions which satisfy both of these equations will represent values of polarization which will be stable in the absence of applied field. Curves corresponding to these equations are plotted in Fig. 18. The straight line, II, represents the equation $P = (1/\beta)F$. The curved lines represent equation I at three different temperatures, T_1, T_2, T_3. For T_3, the only solution is $P = 0$: at this temperature there is no permanent polarization – it is above the critical temperature. For temperature T_1, a solution is given by P_1, which represents the permanent polarization of the material. At this temperature even an infinitesimal applied field will (according to the theory) produce values of P which will in turn produce higher and higher values of F until the stable point of permanent polarization is reached. At temperature T_2 the curves I and II are tangent at the origin: T_2 is the critical temperature below which permanent polarization exists. (In the theory of ferromagnetism T_2 is called the curie point.) To calculate this critical temperature, we set equal the tangents at the origin of the two curves. For low field strengths, the orientation polarizability is so much greater than the distortion polarizability, that as a good approximation we may represent the entire polarization as $n\alpha_0 F$. This orientation polarizability, α_0, is given for low field strengths by the approximation of the Langevin function, $\mu^2/3kT$. Consequently, the equation corresponding to curve I(T_2) at the origin is

$$P = \frac{n\mu^2}{3kT} F .$$

Curve II is represented by $P = (1/\beta)F$; and the equality is written:

$$\frac{n\mu^2}{3kT_c} = \frac{1}{\beta} .$$

If β is $4\pi/3$, as predicted by the Clausius-Mossotti formula for the case of cubic symmetry, then the critical temperature, T_c, for water is calculated to be 1100° K.

A more serious discrepancy is found when the Clausius-Mossotti equation is used to calculate the dielectric constant of water at room temperature. From measurements made in the gaseous state, the water molecule is found to have a permanent moment, μ, of 1.84×10^{-18} esu-cm. The orientation polarizability, $\alpha_0 = \mu^2/3kT$, is thus about 3×10^{-23} cm^3 at room temperature, and the density, n, is about $1/3 \times 10^{23}$ cm^{-3}. Accordingly, the quantity

$$\frac{4\pi}{3} n\alpha$$

is calculated to be around 4; (distortion polarizability is neglected because it is relatively small.) The result is that a negative dielectric constant is calculated from the Clausius-Mossotti formula.

References p. 149

The difficulty is that the Clausius-Mossotti equation cannot be applied to the case of molecules with permanent dipole moments when they are in condensed phase. To do so would be to assume that for each molecule the resultant dipole moment is made up of an induced moment plus a permanent moment which is fully aligned in the direction of the applied field and is equal to the average permanent moment. As we shall see later, it is necessary instead to calculate the microscopic field for each molecule (with its permanent moment pointing in any arbitrary direction), and afterwards take a weighted average based on a Boltzmann distribution.

Although the Clausius-Mossotti equation has limitations, it has a sometimes overlooked value as a 'yardstick' with which to compare experimental results. If the experimental results agree with the theoretical predictions, it can be inferred that the Clausius-Mossotti assumptions adequately characterize the system; if there is a discrepancy, it is apparent that other forces must be present in that particular condensed phase.

The Onsager Equation

A further improvement in the estimation of the microscopic field was introduced by Onsager[2], who included in his calculations all the long-range electrical dipole forces, permanent as well as induced. However, he did not consider the short-range London or chemical bond forces which also tend to orient the molecules.

In following the development of Onsager's equation, we shall relate dielectric constant to the microscopic quantities of permanent dipole moment μ and polarizability α. We know that $\varepsilon - 1 = 4\pi P/E$, where P is the total dipole moment per unit volume; and we know that $P = nm_{av}$, where m_{av} is the average value for dipole moment. It is this quantity, m_{av}, that depends on the actual microscopic field acting on each dipole, and we shall calculate its value in terms of μ, α, and E.

The picture to be borne in mind throughout the discussion of Onsager's equation is the following. Each molecule is located at the center of a spherical hole of radius a, defined such that n holes occupy a volume of one cubic centimeter; *i.e.*, $(4\pi/3)a^3n = 1$, where n is the number of molecules per cubic centimeter. A particular reference molecule, located at the origin of coordinates, has a permanent dipole moment, μ, and a deformation polarizability, α. The assumption is made that the material outside the sphere is homogeneous and has a macroscopic dielectric constant ε. The field inside the hole may be considered as the superposition of two independent fields, one due to the external applied field and the other due to the polarization of the medium by the included dipole: the 'reaction field'.

The first of these, which is the field inside a spherical cavity in a homogeneous dielectric, is obtained by a purely electrostatic calculation. Consider the empty spherical hole in the medium of dielectric constant ε (see Fig. 19). Let the outer region be designated as I, and the region of the hole, II.

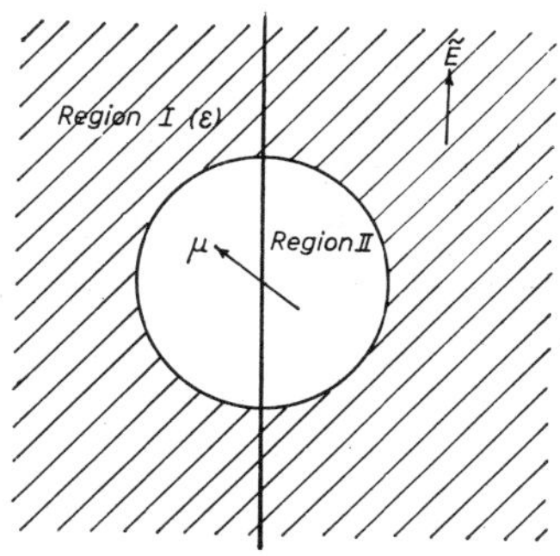

Fig. 19

Rather than trying to solve Laplace's equation (which will always yield the proper answer), we shall make a guess as to the form of the potential. Being guided by the solution to the conducting sphere problem*, we assume that the potential Φ_{I} in region I is due to (1) the external applied field E, and (2) a dipole of unknown strength A located at the origin of coordinates at the sphere center:

$$\Phi_{\mathrm{I}} = - Er \cos \vartheta + \frac{A}{r^2} \cos \vartheta ,$$

r being the distance as measured from the origin, and ϑ being the polar angle. Since no charges are present inside the spherical hole II, we shall assume for this region the simplest condition, a uniform field of unknown strength B:

$$\Phi_{\mathrm{II}} = - Br \cos \vartheta .$$

* See the Appendix of Chapter 1, p. 33 et seq.

If these assumptions are correct, we shall find that A and B are constants depending only on the external field E and the dielectric constant ε. There are two boundary conditions to be satisfied. First, the potential must be continuous at the surface between I and II, since otherwise the field strength there would be infinite. Second, because of the absence of free charge, the normal component of the displacement D must be equal on both sides of the surface in its immediate vicinity*. The normal component of the displacement, D_n, is written in terms of the normal component of field strength E_n; and E_n can be expressed as the partial derivative of potential along the radius. Expressed mathematically, the boundary conditions are:

$$\Phi_{\mathrm{I}} = \Phi_{\mathrm{II}} \text{ at } r = a \; ;$$

$$-\varepsilon \frac{\partial \Phi_I}{\partial r} = - \frac{\partial \Phi_{\mathrm{II}}}{\partial r} \text{ at } r = a \, .$$

Upon substituting, we find:

$$- Ea + A/a^2 = - Ba \; ;$$
$$\varepsilon(E + 2A/a^3) = B \, .$$

We see that A and B really are independent of r and ϑ, and so the form we assumed for potential must be correct. Solving the simultaneous equations, we find that B, which represents the internal field (due to the external applied field) is:

$$B = \frac{3\varepsilon}{2\varepsilon + 1} E \, .$$

This shows that the field is greater within the hole than it is in the dielectric material: effectively, the sphere surface has been polarized by the external field E.

The 'reaction field' is calculated in a very similar manner. We remove the external field and place a permanent polarizable dipole at the origin of coordinates (see Fig. 20). This central dipole has a moment, $\tilde{m}$, of unknown strength, which is the vector sum of the permanent and induced dipole moments. The polar axis is chosen, for convenience, to be along the axis of the dipole $\tilde{m}$. Inside the sphere, in region II, we guess (on the

---

* See the Appendix for further discussion of this point (p. 145).

basis of our previous result) that the total potential Φ_{II} has for its components a dipole potential arising from the total moment, $\tilde{m}$, and a potential arising from a uniform reaction field. In the outside region I we

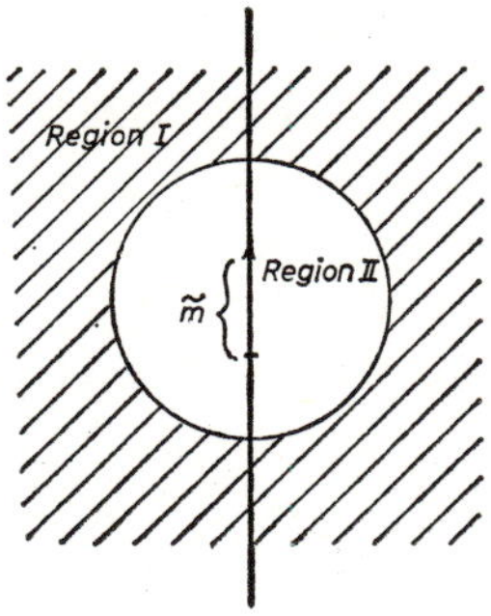

Fig. 20

assume the presence of a pure dipole field of strength $A \cos \vartheta / r^2$, arising from $\tilde{m}$ and the reaction field. In symbols:

$$\Phi_{II} = \frac{m \cos \vartheta}{r^2} - Br \cos \vartheta \; ;$$

$$\Phi_{I} = \frac{A \cos \vartheta}{r^2} \, .$$

Upon introducing the same boundary conditions as before, we find both A and B to be again independent of r and ϑ, implying that the guess about the form of potential was correct. In particular, we calculate the reaction field to be:

$$\tilde{B} = 2 \frac{\varepsilon - 1}{2\varepsilon + 1} \frac{\tilde{m}}{a^3} \, .$$

It may be noted that this reaction field, although caused by the dipole within the sphere, is considered to be an external field. The reason is that although it exists because of the internal dipole, it actually arises from the resultant polarization of dipoles in the external region (I).

The total microscopic field $\tilde{F}$ acting on the dipole $\tilde{m}$ is, then, the sum:

$$\tilde{F} = \frac{3\varepsilon}{2\varepsilon + 1}\,\tilde{E} + 2\,\frac{\varepsilon - 1}{2\varepsilon + 1}\,\frac{\tilde{m}}{a^3}\,,$$

which reduces to $\tilde{E}$, as expected, when $\varepsilon = 1$ or, physically speaking, when the entire medium is vacuum except for the one central dipole. Now, the moment $\tilde{m}$ may be written:

$$\tilde{m} = \tilde{\mu} + \alpha\tilde{F}\,,$$

α being deformation polarizability. The two equations may be combined and solved for $\tilde{m}$ to yield:

$$\tilde{m} = \left[\frac{1}{1 - q(2\alpha/a^3)}\right]\tilde{\mu} + \left[\frac{\alpha p}{1 - q(2\alpha/a^3)}\right]\tilde{E}\,,$$

where for simplicity q is used to represent $(\varepsilon - 1)(2\varepsilon + 1)$ and p represents $3\varepsilon/(2\varepsilon + 1)$. (It should be noted that in general the directions of $\tilde{m}$ and $\tilde{\mu}$ are different.) In like manner the field $\tilde{F}$ is then found to be:

$$\tilde{F} = \frac{1}{1 - q(2\alpha/a^3)}\left[\frac{2}{a^3}\,q\tilde{\mu} + p\tilde{E}\right].$$

Now we are able to calculate the quantity m_{av}, the average dipole moment. Since we are considering an isotropic medium, the average moment will lie in the field direction. It can be calculated by the usual statistical method based on the idea that the potential energy of a dipole in a particular orientation determines the population of dipoles in that direction. (This point is discussed in Chapter 2.) According to this method, m_{av} is calculated by performing the straightforward integration: $m_{av} = C\int m \cos\vartheta\, e^{\tilde{m}\cdot\tilde{F}/kT}\, d\Omega$. However, a simplification can be introduced through an alternate expression for average moment:

$$m_{av} = \frac{1}{1 - q(2\alpha/a^3)}\,\mu_{av} + \frac{\alpha p}{1 - q(2\alpha/a^3)}\,E\,.$$

The second term does not need to be averaged since the distortion polarization is always in the field direction. It is only the expression for

μ_{av}, then, that must be obtained; this is:

$$\mu_{av} = \frac{\mu \int \cos \vartheta \, e^{\tilde{\mu} \cdot \tilde{F}/kT} \, d\Omega}{\int e^{\tilde{\mu} \cdot \tilde{F}/kT} \, d\Omega}.$$

The quantity $\tilde{\mu} \cdot \tilde{F}$ is equal to

$$\frac{1}{1 - q(2\alpha/a^3)} \left[\frac{2q}{a^3} \mu^2 + p\mu E \cos \vartheta \right],$$

where ϑ is the angle between $\tilde{\mu}$ and $\tilde{E}$. The first term of this expression is independent of the field direction, so it can have no effect on the distribution of population and may be taken outside the integral sign. The equation is then:

$$\frac{\mu_{av}}{\mu} = \frac{\int \cos \vartheta \exp \left(\frac{pE}{1 - q(2\alpha/a^3)} \frac{\mu \cos \vartheta}{kT} \right) d\Omega}{\int \exp \left(\frac{pE}{1 - q(2\alpha/a^3)} \frac{\mu \cos \vartheta}{kT} \right) d\Omega},$$

which is the Langevin function (discussed in Chapter 1). Using the approximation for the Langevin function which is valid for ordinary laboratory fields, we find:

$$\mu_{av} = \frac{1}{3} \frac{\mu^2}{kT} \frac{p}{1 - q(2\alpha/a^3)} E.$$

Finally, we obtain the average moment in the form:

$$m_{av} = \frac{p}{1 - q(2\alpha/a^3)} \left(\frac{1}{1 - q(2\alpha/a^3)} \frac{\mu^2}{3kT} + \alpha \right) E.$$

The sphere radius, a, can be eliminated by substituting for $1/a^3$ the equivalent quantity

$$\frac{4\pi}{3} n.$$

This evaluation of m_{av} is the fundamental factor in the Onsager equation. The equation itself is derived from the general expression $\varepsilon - 1 = 4\pi P/E$, which, as we have pointed out, can be expressed as $\varepsilon - 1 = 4\pi n m_{av}/E$.

Before proceeding with the development of the usual form of the Onsager equation, let us consider the expression for m_{av} in terms of previous theory. It can be seen by inspection that in the absence of a permanent moment, the expression for m_{av} reduces to:

$$m_{av} = \frac{p\alpha}{1 - q(2\alpha/a^3)} E ,$$

which yields:

$$\varepsilon - 1 = 4\pi n \frac{p}{1 - q(2\alpha/a^3)} \alpha .$$

Upon substitution of the values of p, q, and a, this becomes:

$$\frac{\varepsilon - 1}{\varepsilon + 2} = \frac{4\pi}{3} n\alpha .$$

the familiar form of the Clausius-Mossotti equation. Thus we see that the Onsager equation includes the Clausius-Mossotti as a special case. This shows in a simple mathematical way that the Clausius-Mossotti theory does not correctly take into account the effect of permanent dipoles.

The difference between the two equations can be seen more specifically by examining the 'correction' factors applied to polarizability. We see that in the Clausius-Mossotti equation (in the form given above involving p and q), total polarizability (designated as α) has been 'corrected' by inclusion of the factor

$$\frac{1}{1 - q2(\alpha/a^3)} .$$

The Onsager equation includes this same factor for the deformation polarizability α, but has an additional factor multiplying the orientation polarizability $\mu^2/3kT$; this may be seen most easily by examining the Onsager expression for m_{av}.

The Onsager equation, in the form we have indicated, relates dielectric constant, ε, deformation polarizability, α, permanent moment, μ, and the number of molecules per cubic centimeter, n. The equation can be made more useful by expressing the unknown quantity α in terms of a measurable parameter, refractive index. Since at optical frequencies the permanent dipole moment cannot follow the rapid changes in the field, we can set the dielectric constant equal to r^2, the square of the index of refraction, and use the expression

$$\frac{r^2 - 1}{r^2 + 2} = \frac{4\pi}{3} n\alpha$$

to find α. When this evaluation of α is substituted into the expression for average moment, we derive the expression:

$$\varepsilon - 1 = 4\pi n \frac{\mu^2}{3kT} \frac{\varepsilon(2\varepsilon + 1)}{3} \left(\frac{r^2 + 2}{2\varepsilon + r^2}\right)^2 + \frac{3\varepsilon(r^2 - 1)}{2\varepsilon + r^2} .$$

Subtracting $r^2 - 1$ from both sides and rearranging terms, we obtain an expression in terms of the quantity $(\varepsilon - r^2)$, which is the difference between the dielectric constants at static and optical frequencies. Thus:

$$(\varepsilon - r^2) \frac{2\varepsilon + r^2}{3\varepsilon} = 4\pi n \frac{\mu^2}{3kT} \left(\frac{r^2 + 2}{3}\right)^2 .$$

This is the Onsager equation. We note that this is a quadratic expression in ε: the positive root represents the real value for dielectric constant. Also, we note that in the limiting case of very low density, as in a gas, both the dielectric constant and the square of refractive index are very nearly unity and to a good approximation:

$$\varepsilon - r^2 = 4\pi n \frac{\mu^2}{3kT} = 4\pi n \alpha_0 ,$$

which agrees with the original equation derived with the assumption of an actual microscopic field equal to the macroscopic applied field. (According to the original equation, $\varepsilon - 1 = 4\pi n(\alpha_D + \alpha_0)$; and since $4\pi n \alpha_D = r^2 - 1$, the quantity $\varepsilon - r^2$ must equal $4\pi n \alpha_0$.) Thus we may conclude that Onsager's results support the validity of the original expression when this is applied to gases at low concentration.

We can test the effectiveness of the Onsager equation by using it to calculate the dielectric constant of water. Using well established values

References p. 149

for the other parameters, we calculate a dielectric constant of 26; the experimental value is about 80. Although far from exact, this is much better agreement than is found using the Clausius-Mossotti equation, which predicts that a material with such a high dipole moment would have a negative dielectric constant. Another way of assessing the equation is to see what it predicts about permanent polarization. From the form of the equation it can be seen that no matter how large the value of the dipole moment and consequently the entire right-hand side of the equation, the dielectric constant will remain finite. Therefore, we cannot arrive with this theory (as we did with the Clausius-Mossotti) at the conclusion that it is possible for a substance to exhibit a finite polarization under the influence of an infinitesimal external field. However, we know that there do exist substances that are permanently polarized. This fact, together with the fact that there is only fair agreement between calculated and actual values of the dielectric constant for highly polarized materials such as water, leads us to suspect that there are still other forces to be taken into account.

The Kirkwood Theory

Not taken into account in the Onsager equation are those short-range forces, chemical in nature, that exist between molecules in condensed media. These are the forces which are, to the greatest extent, responsible for the orientation of molecules in crystal lattices, and they tend to produce such orientation in small crystal-like domains in liquids. The fact that heat is required to evaporate liquids (even non-polar liquids such as benzene) is experimental evidence of the existence of an 'association energy' due to these forces. Because of these short-range forces, the orientation of a dipole (irrespective of the influence of an external field) is not random with respect to its immediate neighbors, and in the consideration of the instantaneous dipole at a given point this should be taken into account.

On the basis of very general considerations, Kirkwood [3] modified the Onsager formula to take into account this condition; in place of the quantity μ^2, he introduced the quantity $\mu\mu^*$, where μ^* represents the total dipole moment in a small domain and is the magnitude of the vector sum of the central dipole moment, plus that moment due to its nearest neighbors. Short-range forces are responsible for the value of μ^* being equal to, greater than, or less than the value of μ; estimation of this quantity requires detailed knowledge of the structure of the surroundings. For the general case Kirkwood expressed the relation as:

$$\mu\mu^* = \mu^2 (1 + Z \overline{\cos \gamma}) ,$$

where Z represents the number of nearest neighbors, and $\overline{\cos \gamma}$ is the average of the cosines of the angles between the dipole moment of the central molecule and the dipole moments of its neighbors.

As a measure of the success of the Kirkwood theory we can again calculate the dielectric constant of water. On the basis of the crystal structure of ice, we consider that there are four nearest neighbors oriented tetrahedrally to any central molecule, and so the value of $\overline{\cos \gamma}$ is 1/3. The correction for μ^2 is thus 7/3, and this raises the calculated value of dielectric constant to 63. This compares more favorably with the experimental value of about 80.

For special cases in which potential energy due to short-range forces is much less than that arising from long-range dipole forces, one would expect the Onsager formula to hold very well. Indeed, it does prove to be adequate for many substances. Also, the accuracy of the Onsager formula is found generally to increase with temperature elevation; this seems consistent with the idea of short-range forces because increased thermal motion would tend to disrupt orientation due to such forces. However, if short-range forces are significant, the Onsager formula is not adequate. This appears to be the situation, for example, in the case of material with permanent polarization: the phenomenon cannot be explained by the Onsager theory and is presumably caused entirely by short-range forces.

In summary, we can say that an investigator today has three 'measuring sticks' to help him describe a condensed system in terms of its measured static dielectric constant. If he is able to explain experimental measurements with the use of the simple Clausius-Mossotti expression, he concludes that there is present in his system only distortion polarization. If the Onsager formula is required to explain the experimental results, he concludes that there are permanent dipole moments present. If only the Kirkwood formula is successful, he concludes that short-range forces comparable in magnitude to long-range dipole forces are present in his system.

References p. 149

Reaction of Dipoles in Viscous Media to an Alternating Field

We have already made use of the idea that dielectric constant when measured (as the square of refractive index) at optical frequencies arises from distortion polarization only, whereas measurements performed at low frequencies include effects due to the orientation of permanent moments. The measured dielectric constant can vary appreciably with frequency: for example, water, which is a polar, highly associated liquid, exhibits at low frequencies a dielectric constant of about 80, but at optical frequencies its dielectric constant is only of the order of magnitude of unity. Such results suggest the investigation of the reaction of polar materials to fields of intermediate frequencies. One of the first investigators to undertake such studies was P. Drude [4]. Through wavelength and frequency measurements of standing waves which were produced in wires imbedded in various media, he was able to calculate dielectric constant. Many materials were found to exhibit a dielectric constant that decreased with increasing frequency: Drude called this behavior 'anomalous dispersion' to contrast it with so-called normal dispersion in which ε increases with frequency (for example, the refractive index for glass is greater for blue light than for red).*

Drude also found that at the frequencies where anomalous dispersion occurs, he could not produce standing waves of high amplitude; in other words, he found that the material was absorbing energy in some strange manner – strange because his media always showed very low conductivity. Today it is generally accepted that it is the interaction of permanent dipoles with the applied field that causes absorption of energy in certain frequency ranges. A reasonable explanation of the phenomenon (for the case of dipoles immersed in a viscous medium) is that the dipoles cause dissipation of energy through friction while attempting to follow the field direction. The faster the dipoles moved, the greater would be the absorption of energy. However, if the frequency were increased beyond the point where the dipoles could follow the field completely, the energy absorption would decrease and so would the measured dielectric constant; at sufficiently high frequencies the dipoles would be unable to orient at all. It would be expected that there would be some characteristic time

---

* Cauchy's discussion of normal dispersion in terms of classical resonators is considered at the end of this chapter.

required for the random dipoles to become oriented in an electric field, and that likewise a particular length of time would characterize the return to random orientation upon removal of the field: this 'relaxation time' would be dependent on viscosity of the medium as well as on the properties of the dipole. A dipole would absorb the greatest amount of energy from a field whose period equals the relaxation time. Thus the relaxation time for dipoles in viscous liquids would have a significançe corresponding to the 'proper frequency of vibration' in a mechanical system. The process of dipoles reverting to random distribution is a diffusion phenomenon, and so the ideas of diffusion theory should be applicable. For this reason we shall next review some aspects of the phenomenon of diffusion.

There are two methods of developing diffusion theory. With one of the methods the results are obtained quickly but the reasoning is not quite sound. The line of reasoning is as follows. Consider a solution containing n molecules of solute per cubic centimeter, and within this liquid consider a rectangular parallelepiped with base area equal to one square centimeter and with height dx. Then from the Van 't Hoff relationship for osmotic pressure, $P = nkT$, the net force in the x direction on the parallelepiped is calculated to be:

$$F = -kT\left(n_0 + \frac{\partial n}{\partial x}\mathrm{d}x\right) + kTn_0 = -kT\frac{\partial n}{\partial x}\mathrm{d}x\,.$$

On a per particle basis, this net force is:

$$f = -\frac{kT}{n}\frac{\partial n}{\partial x}\,.$$

The particle is dragged through the liquid with a velocity v against a resulting frictional force, $-\rho v$, which at equilibrium is equal and opposite to the force f so that $\rho v = f$. (Here ρ is a constant related to viscosity.) The current of particles through one square centimeter is nv, which may be written as fn/ρ. If we substitute, for the force f, the quantity obtained through Van 't Hoff's law, we find that the current in the x direction, J_x, is equal to

$$-\frac{n}{\rho}\frac{kT}{n}\frac{\partial n}{\partial x}\,,$$

so that:

$$J_x = -\frac{kT}{\rho}\frac{\partial n}{\partial x}.$$

But we know that the current through a one square centimeter area which lies perpendicular to a given direction x is:

$$J_x = -D\frac{\partial n}{\partial x},$$

where D is the experimentally measured diffusion constant (here the negative sign indicates that particles flow from a region of high concentration to one of lower concentration). Comparison between the theoretical and this phenomenological expression shows that:

$$D = \frac{kT}{\rho}.$$

Although this is the correct evaluation of D, it was incorrectly derived because it was obtained through force considerations even though no forces act on the molecules in the usual mechanical sense. Just as a perfect gas exerts pressure because of its tendency to increase its entropy, so for this reason and not because of the existence of mechanical force does the diffusion of molecules from higher to lower concentration take place.

Diffusion theory based on a sound line of reasoning was set forth by Einstein [5]. In following his development, let us begin by considering a volume element $\mathrm{d}S$ which is part of the solution containing n particles per cubic centimeter. Throughout the liquid the particles, which may even be colloids, are undergoing Brownian motion because of the constant bombardment by solvent molecules. A change in the number of particles contained within the volume element $\mathrm{d}S$ (which we can designate as $\delta[n\,\mathrm{d}S]$) is accordingly determined by the number of particles which enter or leave the volume because of Brownian motion (δ_1 and δ_2), as well as by the number of particles which enter or leave because of external forces (δ_3). Such an external force could be, for example, an electric field acting on ions in solution: in such a case the ions would describe Brownian motion about their average drift in a direction determined by the field.

However, for the moment we shall assume that external forces are absent, so that:

$$\delta(n\,\mathrm{d}S) = \delta_1 + \delta_2\,.$$

Now suppose that there has elapsed since the starting time, a time which is so long compared with collision time that all the original particles within $\mathrm{d}S$ have left, but yet which is short enough so that only the immediate surroundings of $\mathrm{d}S$ contribute particles to it. The mathematical description of this situation is:

$$\delta(n\,\mathrm{d}S) = \delta_1 + \delta_2 = -\,n\,\mathrm{d}S + \int_{\mathrm{all}\,S'} n'\,\mathrm{d}S'\,\varphi\,\mathrm{d}S\,,$$

where the prime indicates, for volume and concentration, the immediate neighborhood. The first term, $-\,n\,\mathrm{d}S$, represents the depletion of the original particles. The second term represents the influx of particles to $\mathrm{d}S$ from the surroundings: $\varphi\,\mathrm{d}S$ is the probability that one particle originally in $\mathrm{d}S'$ will arrive in $\mathrm{d}S$ after time τ; multiplication by $n'\,\mathrm{d}S'$ gives the total number of particles contributed by the volume element $\mathrm{d}S'$. The function φ depends on the distance between $\mathrm{d}S$ and $\mathrm{d}S'$ and on time τ. Now, the volume $\mathrm{d}S$, being constant, may be removed from under the integral sign, leaving the expression $\mathrm{d}S\int_{\mathrm{all}\,S'}\varphi n'\,\mathrm{d}S'$. Also, if the coordinates of $\mathrm{d}S$ are given by x, y, z, we may write for the coordinates of $\mathrm{d}S'$, $x+\xi$, $y+\eta$, $z+\zeta$. Because only the immediate neighborhood is considered, a sufficiently good approximation to n' is given by:

$$n' = n + \xi\frac{\partial n}{\partial x} + \eta\frac{\partial n}{\partial y} + \zeta\frac{\partial n}{\partial z}$$

$$+\frac{1}{2}\left(\xi^2\frac{\partial^2 n}{\partial x^2} + \eta^2\frac{\partial^2 n}{\partial y^2} + \zeta^2\frac{\partial^2 n}{\partial z^2} + 2\xi\eta\frac{\partial^2 n}{\partial x\partial y} + 2\xi\zeta\frac{\partial^2 n}{\partial x\partial z} + 2\eta\zeta\frac{\partial^2 n}{\partial y\partial z}\right).$$

Substituting this expression for n', let us consider the integral term by term. The first term is $\int_{\mathrm{all}\,S'} n\varphi\,\mathrm{d}S'$, which, in accordance with the definition of φ, is just n. Integration of the remaining expression yields terms involving the averages $\overline{\xi}$, $\overline{\eta}$, $\overline{\zeta}$, $\overline{\xi\eta}$, $\overline{\xi\zeta}$, $\overline{\eta\zeta}$, $\overline{\xi^2}$, $\overline{\eta^2}$, and $\overline{\zeta^2}$. The quantities $\overline{\xi}$, $\overline{\eta}$, and $\overline{\zeta}$, are 0, since they have been obtained by integration, over all space, of odd functions. (Physically speaking, particles move backwards or forwards with equal probability.) Likewise, the mixed averages $\overline{\xi\eta}$, $\overline{\xi\zeta}$, and $\overline{\eta\zeta}$ are also zero. The expression that remains is:

$$\delta(n\,\mathrm{d}S) = -\,n\,\mathrm{d}S + n\,\mathrm{d}S + \frac{1}{2}\left(\overline{\xi^2}\frac{\partial^2 n}{\partial x^2} + \overline{\eta^2}\frac{\partial^2 n}{\partial y^2} + \overline{\zeta^2}\frac{\partial^2 n}{\partial z^2}\right)\mathrm{d}S\,.$$

Since $\delta(n\,\mathrm{d}S)$ may be written as

$$\tau\frac{\partial n}{\partial t}\,\mathrm{d}S\,,$$

and since there is no preferred direction of motion, so that $\overline{\xi^2} = \overline{\eta^2} = \overline{\zeta^2}$, then we may write the equation as:

$$\tau\frac{\partial n}{\partial t}\,\mathrm{d}S = \frac{\overline{\xi^2}}{2}\left(\frac{\partial^2 n}{\partial x^2} + \frac{\partial^2 n}{\partial y^2} + \frac{\partial^2 n}{\partial z^2}\right)\mathrm{d}S\,.$$

Dividing by τ, we obtain the equation:

$$\frac{\partial n}{\partial t} = \frac{\overline{\xi^2}}{2\tau}\left(\frac{\partial^2 n}{\partial x^2} + \frac{\partial^2 n}{\partial y^2} + \frac{\partial^2 n}{\partial z^2}\right).$$

Another equation for change in concentration with time may be written in terms of diffusion constant. Recalling that the current of particles per second through a unit area perpendicular to the x axis is $-\,D(\partial n/\partial x)$, we see that in time τ the change in the number of particles due to a current in the x direction in a parallelepiped with base area of unity and height $\mathrm{d}x$ is:

$$\left[D\frac{\partial n}{\partial x}\tau + \frac{\partial}{\partial x}\left(D\frac{\partial n}{\partial x}\tau\right)\mathrm{d}x\right] - D\frac{\partial n}{\partial x}\tau\,,$$

or, briefly:

$$D\frac{\partial^2 n}{\partial x^2}\tau\,\mathrm{d}x\,.$$

But the change in the number of particles in the parallelepiped (which has a base area of unity) is

$$\frac{\partial n}{\partial t}\tau\,\mathrm{d}x\,.$$

Together these two expressions yield the formula for diffusion in one dimension:

$$D\frac{\partial^2 n}{\partial x^2} = \frac{\partial n}{\partial t}.$$

The expression for three dimensions is obtained through the same generalization used in the Appendix of Chapter 1 in developing the expression for the divergence of a vector field. The resulting expression is the diffusion equation:

$$\frac{\partial n}{\partial t} = D\left(\frac{\partial^2 n}{\partial x^2} + \frac{\partial^2 n}{\partial y^2} + \frac{\partial^2 n}{\partial z^2}\right).$$

In other words, the current of particles is established by the gradient of concentration, and the resultant change with time of the concentration is given by the divergence of this current.

Comparison of the diffusion equation with Einstein's formula shows that the constant D represents the average, $\overline{\xi^2}/2\tau$. This indicates that $\overline{\xi^2}$ is a linear function of time, as it also turns out to be in other random motion problems. Thus the Einstein theory provides a physical interpretation of the experimental constant D. For example, for a sugar solution the measured diffusion constant is 1 cm^2/day, or 10^{-5} cm^2/sec; this means that for one second the root mean square of x displacement, $\sqrt{\overline{\xi^2}}$, is 0.4×10^{-2} cm, an astonishingly small fraction of the average distance covered by a molecule of the same mass at the same temperature in the highly rarified gaseous state where molecular paths are long between each collision. The task of checking the constancy of $\overline{\xi^2}/2\tau$ was undertaken by Perrin[6], who observed colloidal particles under the microscope and found that within experimental error this quantity was really constant.

This development has established the microscopic relationship between diffusion and Brownian motion. Now we can consider the influence of external forces. Suppose that an external force F (such as gravity or an electric field) has been applied, and that particles are moving with a drift velocity v, superimposed on the Brownian motion, of magnitude $v = F/\rho$, where ρ is the friction constant. The drift current in the x direction due to the external force, J_{xF}, is therefore equal to nF/ρ. The total current in the

References p. 149

x direction, arising from the sum of drift current and Brownian motion, is:

$$J_x = -D\frac{\partial n}{\partial x} + \frac{nF}{\rho};$$

and the total timewise change in the number of particles per unit volume is (for one dimension):

$$\left(\frac{\partial n}{\partial t}\right)_x = -\frac{\partial}{\partial x}\left(-D\frac{\partial n}{\partial x} + \frac{nF}{\rho}\right).$$

This formula should be a good approximation even for applied forces that change with time, so long as the period of the applied force is long compared with collision time. We can estimate collision time by means of the kinetic energy expression. If we take for the average translational kinetic energy, $mv^2/2$, of a molecule the value $3\,kT/2$, then

$$v = \left(\frac{3\,kT}{m}\right)^{1/2}.$$

Average collision time is calculated as the time required for a molecule to travel the average intermolecular distance. For liquid water, as an example, the velocity v is calculated to be 6×10^4 cm/sec; since 18 cubic centimeters of water contain 6×10^{23} molecules, the average distance between them is

$$\left(\frac{18}{6 \times 10^{23}}\right)^{1/3} \text{cm},$$

or about 3Å; and the time required to move this distance is $1/2 \times 10^{-12}$ sec. Accordingly, for the case of water the period of the external force must be greater than 10^{-12} seconds if our derived formula is to be valid; this restriction corresponds to a lower limit for wavelength of 300 microns, which lies in the infrared region.

The development for one dimension, given above, yielded a formula which may be expected to hold in the general case. However, it is possible to evaluate the diffusion constant D, through the application of these ideas to a particular problem. Suppose a solution has been sitting undisturbed in a beaker and has reached equilibrium under the influence of gravity. Some particles will be moving downward under the influence of gravity, others will be moving upward because of Brownian motion, but there will be no total current; *i.e.*,

$$J_T = -D\frac{\partial n}{\partial x} + \frac{nF}{\rho} = 0\,.$$

The gravity force F is associated with a potential energy V according to the usual expression, $F = -\partial V/\partial x$; thus

$$J_T = -D\frac{\partial n}{\partial x} - \frac{n}{\rho}\frac{\partial V}{\partial x} = 0\,,$$

which in integrated form is $n = A\,\mathrm{e}^{-V/\rho D}$. But we know from Boltzmann's law that $n = A\,\mathrm{e}^{-V/kT}$. Since both the diffusion law and Boltzmann's law have been judged to be sound physical laws on the basis of experimental results that have accrued throughout many years, we can conclude that ρD is identically equal to kT. Thus the diffusion constant is:

$$D = kT/\rho\,.$$

This is the same result that we obtained earlier through somewhat faulty reasoning involving consideration of forces.

Substituting kT/ρ for D in the diffusion equation, we obtain for one dimension:

$$\frac{\partial n}{\partial t} = -\frac{\partial}{\partial x}\left(-\frac{kT}{\rho}\frac{\partial n}{\partial x} + \frac{nF}{\rho}\right),$$

or, alternately:

$$\frac{\rho}{kT}\frac{\partial n}{\partial t} = -\frac{\partial}{\partial x}\left(-\frac{\partial n}{\partial x} + \frac{nF}{kT}\right).$$

This latter expression may be used in connection with ion mobility experiments to determine ρ and thus also diffusion constant D.

The three-dimensional diffusion equation, in the absence of external forces, shows a resemblance to Poisson's equation for the electrostatic field; in both equations there appears the Laplacian operator, usually designated as $\triangle$. In Poisson's equation as applied to the case of a region containing no charge, the Laplacian of the potential, $\triangle\Phi$, is zero; this electrostatic equation is discussed in Chapter 1. There is one solution to this equation that depends only on the distance r and not on the polar angles, and which becomes infinite as the point $r = 0$ is approached. This solution represents the physical situation of potential arising from charge concentrated in a 'point' located at $r = 0$, and having the value $\Phi = 1/r$. This solution to the electrostatic equation suggests that there may also be a spherically symmetrical solution to the diffusion equation: the Laplacian would then be

$$\frac{d^2}{dr^2} + \frac{2}{r}\frac{d}{dr},$$

which becomes infinite at zero time at a point $r = 0$. It can be verified by substitution into the differential equation that such a solution is:

$$n = c\,\frac{e^{-r^2/4Dt}}{t^{3/2}}\,.$$

The physical implication of this formula becomes apparent as vanishingly small times are considered: the number of particles per unit volume reaches higher and sharper maxima at $r = 0$, implying that at the initial instant all the particles were gathered there. Fig. 21 illustrates the appearance of the function for increasing values of the time; the spread of particles through diffusion away from their initial concentration as

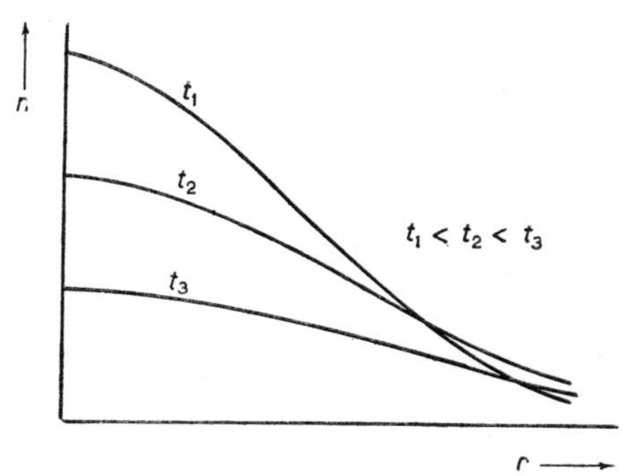

Fig. 21

time goes by is readily apparent. The constant c is determined by normalizing the expression; *i.e.*, by setting

$$\int_0^\infty 4\pi r^2 n \, \mathrm{d}r = 1 \, .$$

The final normalized expression is:

$$n = \frac{\mathrm{e}^{-r^2/4Dt}}{(4\pi Dt)^{3/2}} \, .$$

The normalization allows one to consider just one individual particle, for the normalized formula represents the probability that after a time t a particular particle will be at a distance r from its starting point. Consequently, this form of the equation is very useful. As a very simple example, suppose a particle is located at a point q, and we wish to estimate the probability that this particle will be located in the volume element $\mathrm{d}S$ at a distance r from the point q after a time t has elapsed. The desired probability is $P\mathrm{d}S$, where P is the probability expressed as the normalized function n and represents the probability of finding the particle in a unit volume at the given location. As another example, the dependence on time of the average distance, r_{av}, traveled away from a given starting point for particles in a medium with an associated diffusion constant D can be found through the probability expression. The expression of interest is

$$r_{av} = \int_0^\infty rP \, \mathrm{d}S \, ;$$

this equals

$$\frac{4}{\sqrt{\pi}} (Dt)^{1/2} \, ,$$

as may be easily verified. This square root dependence of time is the logical consequence of the assumption that $\overline{\xi}$, $\overline{\eta}$, and $\overline{\zeta}$, are zero in the microscopic picture of diffusion, which led in turn to the conclusion that $\overline{\xi^2}$ is a linear function of time (for the very short time τ). As a final example, we might calculate the probability that a particle will return after a time t to the unit volume from which it started at $t = 0$. The answer is obtained by setting the distance r equal to zero in the normalized probability expression for n. The result is proportional to $1/t^{3/2}$.

References p. 149

Now that the foundation of diffusion theory has been laid, we are ready to consider how a surrounding viscous medium affects the ability of polar molecules to orient themselves with a time-varying field. Following the line of reasoning used in Einstein's treatment of Brownian motion, we can develop the concept of 'rotary diffusion'.

Let us consider a spherical particle in the medium. Let the function $f(\vartheta, \varphi)\, d\Omega$ represent the number of particles at any location within the medium which point in the direction of the solid angle $d\Omega$, the orientation of $d\Omega$ being given by the angles ϑ and φ of the usual polar coordinates. The direction of a particle will be described by the direction of its permanent electrical moment. The distribution $f(\vartheta, \varphi)$ changes with time because of the influence of (a) Brownian motion, and (b) external forces. Let us first consider cause (b). In the presence of a uniform electric field E directed along the polar axis, each molecule (of moment μ) experiences a torque M of magnitude $-\mu E \sin \vartheta$, the minus sign indicating that the torque tends to decrease the angle ϑ. Under the influence of the torque M (neglecting angular acceleration), the dipole rotates with an angular velocity ω given by $\zeta\omega = M$, where ζ is the friction constant, or the torque per unit angular velocity *. If the molecule were really a small sphere of radius a, and the viscosity acting on the sphere were the macroscopic viscosity η of the medium, then ζ could be calculated using the well known Stokes relation, $\zeta = 8\pi\eta\, a^3$, for the rotation of a sphere. One may feel confident that this assumption will yield a good approximation to the true state of affairs because of the existence of experimental verification for the linear case. For this case Stokes calculated that the friction constant ρ for a non-rotating sphere of radius a moving in a straight line with velocity v through a fluid is represented by the expression $\rho = 6\pi\eta\, a$; the quantity ρv is the frictional force. If the linear friction constant for a molecule in water is calculated by this formula, with η taken as the macroscopic viscosity of water, the predicted mobility in water very nearly equals that measured experimentally. In view of the agreement for the linear case, we shall take ζ to be $8\pi\eta\, a^3$ for the rotary sphere.

Next let us consider the effect of Brownian motion. The method of calculation of the time rate of change of the angular distribution is similar to the method used for the case of ordinary diffusion; a detailed

* This is superimposed on the Brownian motion.

presentation may be found in the literature[7]. Essentially, the procedure is to express the distribution function around the region of interest, $d\Omega$, as a Taylor series in polar rather than Cartesian coordinates, with just first and second degree terms retained. As in the theory of ordinary diffusion, the assumption is made that diffusion time is long compared with collision times, but short enough so that the distribution function will be influenced by the rotation of only those particles whose directions lie in the immediate 'neighborhood' of $d\Omega$.

The whole expression, taking into account both Brownian motion and the applied field, is:

$$\frac{\partial f}{\partial t} = \frac{1}{\sin \vartheta} \frac{\partial}{\partial \vartheta} \left[\sin \vartheta \left(\frac{\overline{\vartheta^2}}{4\tau} \frac{\partial f}{\partial \vartheta} - \frac{Mf}{\zeta} \right) \right] \; ;$$

$\overline{\vartheta^2}$ represents the average of ϑ^2 over a time τ, and from considerations of the symmetry of the problem (the applied field is parallel to the polar axis), we know that f is a function of the polar angle ϑ only. A study of the assumptions in regard to the time τ leads one to expect that this expression should hold even for time-dependent fields, so long as the torques act for times that are long compared to collision times.

The term $\overline{\vartheta^2}/4\tau$ may be evaluated by application of these ideas to a particular problem. Suppose a static field has been applied to a medium containing polar molecules for a time sufficiently long so that the distribution function no longer changes with time. This constitutes a state of equilibrium, in which the current of particles vanishes everywhere, and so our rotary diffusion expression for $\partial f/\partial t$ must equal zero. We therefore set

$$\frac{\overline{\vartheta^2}}{4\tau} \frac{\partial f}{\partial \vartheta} - \frac{Mf}{\zeta} = 0 \, .$$

But the torque M which is applied to a dipole that is inclined at the angle ϑ to the field E is equal to $-\mu E \sin \vartheta$, which can be written as

$$\frac{\partial}{\partial \vartheta} (\mu E \cos \vartheta) \, .$$

Consequently, we may write:

$$\frac{\overline{\vartheta^2}}{4\tau} \frac{\partial f}{\partial \vartheta} - \frac{\mu E}{\zeta} f \frac{\partial}{\partial \vartheta} \cos \vartheta = 0 \, .$$

References p. 149

Upon separation of variables and integration of the resulting equation we obtain:

$$f = A \exp \left[\frac{\mu E \cos \vartheta}{\dfrac{\overline{\vartheta^2}}{4\tau} \zeta} \right].$$

We also expect that, because we are considering an equilibrium situation, the application of Boltzmann's law will yield a correct expression for the distribution function. The energy of a dipole of magnitude μ which is oriented at an angle ϑ to an applied field E is $-\mu E \cos \vartheta$; and so according to Boltzmann's law, the distribution function must be proportional to the quantity $e^{\mu E \cos \vartheta / kT}$. Comparison of this relation with the equation for f derived above, shows that

$$\frac{\overline{\vartheta^2}}{4\tau} \zeta$$

stands in the place of kT, and since the two expressions must yield identical results, we conclude that for the case of rotary diffusion:

$$\frac{\overline{\vartheta^2}}{4\tau} = \frac{kT}{\zeta}.$$

Similar considerations applied to the case of ordinary diffusion gave us the corresponding equation:

$$\frac{\overline{\vartheta^2}}{2\tau} = \frac{kT}{\rho} = D.$$

The final form of the rotary diffusion equation, which we write so as to show explicitly the influence of the friction constant ζ, is:

$$\frac{\zeta}{kT} \frac{\partial f}{\partial t} = \frac{1}{\sin \vartheta} \frac{\partial}{\partial \vartheta} \left[\sin \vartheta \left(\frac{\partial f}{\partial \vartheta} - \frac{Mf}{kT} \right) \right].$$

In this connection, too, it was the work of Perrin [8] that provided experimental verification for diffusion theory: his measurements of the rotational motions of colloidal particles agreed with the predictions of the rotary diffusion theory. These results, taken together with his earlier work supporting the validity of the translational diffusion theory, and with his experiments concerning the height distribution of colloidal particles acted upon by gravity, showed conclusively that the behavior of colloidal particles can be described by the same laws that were developed for atoms and molecules. One of the first practical applications of this method was the determination of Avogadro's number. Having measured $\overline{\xi^2}/2\tau$ for a particular spherical colloid at a temperature T, Perrin introduced the friction constant ρ as calculated from Stoke's law ($\rho = 6\pi\eta\, a$) and obtained a value for Boltzmann's constant, k, by means of the equation:

$$k = \frac{\rho}{T}\frac{\overline{\xi^2}}{2\tau}\,.$$

Then, using the well known gas constant R, he calculated Avogadro's number according to the relation, $N = R/k$. Perrin's experiments yielded values for N which lay in a range (between 6 and 7×10^{23} molecules per gram) that includes today's accepted value of 6.023×10^{23}.

Let us develop this mathematical expression further by considering a particular physical situation. Suppose a static field E has been applied for a long time to a solution of polar molecules; the distribution function $f(\vartheta)$ is then represented in accordance with Boltzmann's law as

$$f = A \exp\left(\frac{\mu E}{kT}\cos\vartheta\right).$$

The quantity $\mu E/kT$ is much smaller than unity in most instances, and so we may express the distribution function to good approximation as:

$$f = A\left(1 + \frac{\mu E}{kT}\cos\vartheta\right).$$

Suppose now that at a certain time, which we take as zero, the field is suddenly removed: this means that the torque, M, vanishes, and the equation to be solved becomes:

$$\frac{\zeta}{kT}\frac{\partial f}{\partial t} = \frac{1}{\sin\vartheta}\frac{\partial}{\partial\vartheta}\left(\sin\vartheta\frac{\partial f}{\partial\vartheta}\right),$$

References p. 149

where the function f must begin as a Boltzmann distribution function at time $t = 0$. This diffusion equation will predict that at time greater than zero there will be a decay of orientation. With these considerations in mind, we make a guess that the new distribution function will be of the form:

$$f = A\left[1 + \frac{\mu E}{kT}\cos\vartheta\varphi(t)\right],$$

hoping that φ will turn out to be a function of time only. Upon substitution, we obtain the relations:

$$\frac{\zeta}{kT}\frac{\partial f}{\partial t} = \frac{\zeta}{kT}\frac{\mu E}{kT}A\cos\vartheta\frac{\mathrm{d}\varphi}{\mathrm{d}t};$$

$$\frac{\partial}{\partial\vartheta}\left(\sin\vartheta\frac{\partial f}{\partial\vartheta}\right) = -2A\frac{\mu E}{kT}\sin\vartheta\cos\vartheta\,\varphi(t).$$

Finally:

$$\frac{\zeta}{kT}\frac{\mathrm{d}\varphi}{\mathrm{d}t} = -2\varphi.$$

Upon integration, this equation yields the result:

$$\varphi = c\,\mathrm{e}^{-\frac{2kT}{\zeta}t}.$$

Thus it is shown that φ really is a function of time alone. The constant $\zeta/2kT$ has the dimension of time, and if we call this the 'relaxation time' and give it the symbol τ, we may write:

$$\varphi = c\,\mathrm{e}^{-t/\tau}.$$

Thus, the relaxation time represents that time required for φ to decay to $1/\mathrm{e}$ of its original value. The final solution for the distribution function is:

$$f = A\left(1 + \frac{\mu E}{kT}\cos\vartheta\,\mathrm{e}^{-t/\tau}\right),$$

which represents an exponential decay from the Boltzmann distribution to a uniform distribution.

The relaxation time τ may be incorporated into the basic differential equation which would then appear as:

$$2\tau \frac{\partial f}{\partial t} = \frac{1}{\sin \vartheta} \frac{\partial}{\partial \vartheta} \left[\sin \vartheta \left(\frac{\partial f}{\partial \vartheta} - \frac{Mf}{kT} \right) \right].$$

With the aid of this equation we can discuss quantitatively the dispersion of a polar liquid. First of all, we can gain some idea of the frequency range in which to expect 'anomalous dispersion' to occur by making an estimate of the relaxation time τ, for as we have pointed out, anomalous dispersion occurs in the range around that frequency whose period corresponds to the relaxation time. We know that:

$$\tau = \frac{\zeta}{2kT} = \frac{4\pi\eta a^3}{kT}.$$

For the estimate we will use values corresponding to ordinary molecular radii, viscosity, and temperature: for a particle radius of 3×10^{-8} cm, a viscosity of 0.01 poise, and a value for kT of 4×10^{-14} ergs, we calculate τ to be 0.8×10^{-10} seconds. From this we infer that a system with these parameters would exhibit anomalous dispersion in the centimeter wavelength range.

Now let us proceed toward a more detailed description of dispersion by solving our differential equation for the case of a periodic applied field. The field is represented as $E = E_0 e^{i\omega t}$, where E_0 is the amplitude and ω is the frequency. Therefore the torque M equals $-\mu E_0 \, e^{i\omega t} \sin \vartheta$, and the differential equation becomes:

$$2\tau \frac{\partial f}{\partial t} = \frac{1}{\sin \vartheta} \frac{\partial}{\partial \vartheta} \left[\sin \vartheta \left(\frac{\partial f}{\partial \vartheta} + \frac{\mu E_0}{kT} f e^{i\omega t} \sin \vartheta \right) \right].$$

In attempting to write a solution to this equation, we recall that the distribution function that applies to the case of sudden removal of a steady electric field was obtained from the static solution, Boltzmann's law, by including in the orientation term an exponentially decreasing time factor, $e^{-t/\tau}$. The solution was:

$$f = A \left(1 + \frac{\mu E_0}{kT} \cos \vartheta \, e^{-t/\tau} \right).$$

It is reasonable to expect that the solution corresponding to the application of a sinusoidal field should be of the form:

$$f = A\left(1 + \beta \frac{\mu E_0}{kT} \cos \vartheta \, e^{i\omega t}\right),$$

where β may be some function of the frequency ω, but not of t or ϑ.

When the proposed solution for f is substituted into our differential equation, we obtain one linear and one quadratic term in the time factor. However, the coefficient of the quadratic term contains the quantity $(\mu E_0/kT)^2$, and since this quantity is much smaller than unity for field strengths ordinarily used, it can be neglected in first approximation. Upon substituting and performing the indicated operations, we obtain:

$$2\tau \, i\omega\beta A \frac{\mu E_0}{kT} e^{i\omega t} \cos \vartheta = (1 - \beta) \, 2A \frac{\mu E_0}{kT} e^{i\omega t} \cos \vartheta \, .$$

It should be pointed out that in this derivation we have made use of the exponential representation of a sinusoidal function. Actually, such substitution is valid only when the involved terms are linear in $e^{i\omega t}$. This is because when such a substitution is made, the implication is that at the end of the operation the real part (that representing the physical situation) will be taken, and whereas the real part of $e^{i\omega t}$ is $\cos(\omega t)$, the real part of the product $e^{i\omega t} \cdot e^{i\omega t}$ is $\cos(2\omega t)$ and not $\cos^2(\omega t)$. The reason that the substitution is permissible in our equation is that the quadratic term is small and is dropped.

Solving the equation for β, we obtain:

$$\beta = \frac{1}{1 + i\omega\tau} \, .$$

This result indicates, first of all, that our guess as to the form of f was correct, for β turns out to be a function of frequency alone. This form of frequency dependence is, in a general way, what one might expect. In the limit of zero frequency, β is unity and f reverts to the Boltzmann function. As frequency increases, the absolute value of β decreases, which, interpreted in a physical sense, means that the dipoles become less able to follow the rapid alternation of the field. Finally, in the limit of infinite frequency, β vanishes and the distribution becomes uniform (f is constant).

Substituting our evaluation of β, we obtain for the distribution function when the system is under the influence of a sinusoidal field:

$$f = A\left[1 + \frac{1}{1 + \mathrm{i}\omega\tau}\frac{\mu E_0}{kT}\cos\vartheta\, \mathrm{e}^{\mathrm{i}\omega t}\right].$$

Since our object is to discuss dispersion, the quantity we want to evaluate is dielectric constant, ε. This can be calculated from the average dipole moment in the field direction, m_{av}, which we can express by simple amendment of the expression that applies to the static case. In the static case, for a Boltzmann distribution characterized by the function

$$f = A\left[1 + \frac{\mu E_0}{kT}\cos\vartheta\right],$$

we found that (for the usual small fields),

$$\frac{m_{av}}{\mu} = \frac{1}{3}\frac{\mu E_0}{kT},$$

which is 1/3 of the coefficient of $\cos\vartheta$ appearing in the expression for f. For the case of a time-varying field, this coefficient is

$$\frac{1}{1 + \mathrm{i}\omega\tau}\frac{\mu E_0}{kT}\mathrm{e}^{\mathrm{i}\omega t},$$

and we infer that

$$\frac{m_{av}}{\mu} = \frac{1}{3}\frac{1}{1 + \mathrm{i}\omega\tau}\frac{\mu E_0}{kT};$$

we need not include the term $\mathrm{e}^{\mathrm{i}\omega t}$ here because we are interested only in the phase difference between the applied field and the average moment, not in the over-all time variation.

This expression for m_{av} (which is the average moment for a single particle) involves the microscopic field, and so for the calculation of dielectric constant of the condensed medium, we should use an equation such as the Onsager formula. Of course, the Onsager formula would have to be modified so as to take into account the effect of the time variation of field, and this leads to such computational difficulties that

References p. 149

for a dilute solution of polar molecules we shall use instead the approximate formula $\varepsilon - \nu^2 = 4\pi\, n\, \alpha_0$, which for our case is:

$$\varepsilon - \nu^2 = 4\pi n \frac{\mu^2}{3kT} \frac{1}{1 + \mathrm{i}\omega\tau}.$$

This yields a complex value for ε, which we can accordingly write as $(\varepsilon_1 - \mathrm{i}\varepsilon_2)$, with ε_1 representing the real part and $-\mathrm{i}\varepsilon_2$ representing the imaginary part; ε_2 is a real number and the negative sign is arbitrary. The significance of the complex evaluation can be seen in terms of the relation between applied field E, which equals $E_0\, \mathrm{e}^{\mathrm{i}\omega t}$, and the displacement D, which equals εE or $(\varepsilon_1 - \mathrm{i}\varepsilon_2) E_0{}^{\mathrm{i}\omega t}$. The real part of E, which represents the physical quantity, is $E_0 \cos(\omega t)$, and the real part of D is found to be $\varepsilon_1 E_0 \cos(\omega t) + \varepsilon_2 E_0 \sin(\omega t)$. Whereas E is a cosine function, D turns out to include a sine function, and this indicates that there is a phase difference between the field E and the displacement D.

The part ε_2 of the complex dielectric constant can be shown to be a measure of the energy that is absorbed from the field and dissipated through the friction created by the dipole's attempt to follow the field. This is discussed in detail in the Appendix.

In order to derive expressions for the components ε_1 and ε_2, the quantity

$$\frac{1}{1 + \mathrm{i}\omega\tau},$$

which appears in the expression for the dielectric constant, must be rationalized. This is done by making use of the identity

$$\frac{1}{1 + \mathrm{i}x} \equiv \frac{1}{1 + x^2} - \frac{\mathrm{i}x}{1 + x^2}.$$

Upon separating the real and imaginary parts of the resulting equation, we find:

$$\varepsilon_1 = \nu^2 + \frac{A}{1 + \omega^2\tau^2};$$

$$\varepsilon_2 = \frac{A\omega\tau}{1 + \omega^2\tau^2},$$

where A represents the quantity

$$4\pi n \frac{\mu^2}{3kT}.$$

These functions have the graphical appearance shown in Fig. 22. When plotted on semi-logarithmic paper, the curve for ε_2 is symmetric about the frequency $\omega = 1/\tau$; the greatest rate of decrease of ε_1 also occurs at this frequency. This graphical representation clearly shows τ to be a parameter which is to dipoles in a viscous medium what resonance frequency is to an oscillator. There exist some systems for which these theoretical curve shapes adequately represent experimentally determined values. But usually, one finds that the experimental curve for ε_2 is of the shape shown as a dotted curve in Fig. 23. This result can be explained if

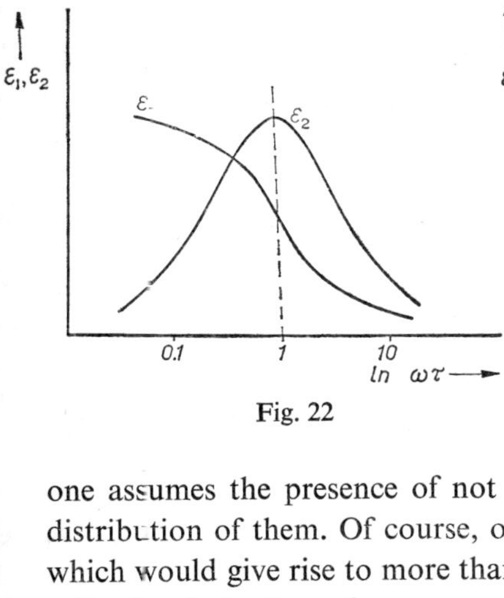

Fig. 22

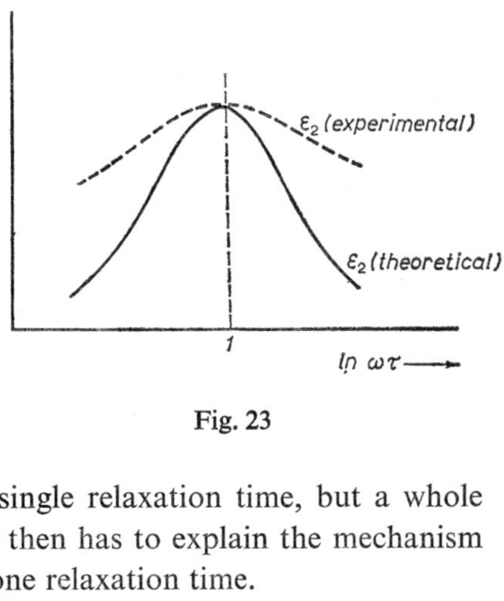

Fig. 23

one assumes the presence of not a single relaxation time, but a whole distribution of them. Of course, one then has to explain the mechanism which would give rise to more than one relaxation time.

In the derivation of our expression for dielectric constant, we used for the sake of simplicity a formula less exact than the Onsager equation. The difficulty in using the Onsager equation is not in the derivation of the general expression: for our case, one simply includes the factor $1/(1 + i\omega\tau)$ in the Onsager expression for average moment, obtaining:

$$(\varepsilon - \nu^2)\frac{2\varepsilon + \nu^2}{3\varepsilon} = \left(\frac{\nu^2 + 2}{3}\right)^2 4\pi n \frac{\mu^2}{3kT}\frac{1}{1 + i\omega\tau}.$$

References p. 149

The difficulty, obviously, is in separating this equation into real and imaginary parts. It is necessary to resort to the method of successive approximations by which one may obtain solutions which become better and better according to the limit of one's patience. The approximation can be made in the following way. The entire range of magnitude of the troublesome term

$$\frac{2\varepsilon + \nu^2}{3\varepsilon}$$

runs from about 2/3 for very low frequency to unity for high frequency. If one substitutes the symbol $1/\beta$ for this expression, the Onsager formula becomes:

$$\varepsilon - \nu^2 = \beta \left(\frac{\nu^2 + 2}{3}\right)^2 4\pi n \frac{\mu^2}{3kT} \frac{1}{1 + i\omega\tau} \equiv \frac{K}{1 + i\omega\tau}.$$

Since β is almost constant, a first approximation may be obtained by introducing its average value; then separation proceeds with ease. In this way one finds for the first approximation step the values

$$\varepsilon_1 = \nu^2 + \frac{K}{1 + \omega^2\tau^2}$$

and

$$\varepsilon_2 = \frac{K}{1 + \omega^2\tau^2}.$$

A new value for β may be calculated on the basis of these values for ε_1 and ε_2, and the whole process repeated as often as desired to obtain closer and closer approximations to the complex dielectric constant ε.

All of this development has been based on the idea that a dipole is a small sphere which is made to rotate in a homogeneous viscous medium, and that the effective viscosity acting to retard the sphere is the same as that which is measured macroscopically. The resulting theory is able, in most cases, to explain the observed phenomena for dipoles in liquid media.

One might ask if the theory is valid for dipoles in the solid or gaseous states. It turns out that, in the case of solids, the viscosity which would be predicted from examination and plotting of data taken from measurements of radio frequency dispersion is much lower than the actual viscosity; to investigate the problem further, one must look into the more recent theories that treat viscosity in terms of the rotations of molecules over energy barriers in the medium. These aspects will not be discussed in this book. The theory as presented is also insufficient for the treatment of polar gases, and for this we shall use another approach.

Polar Gases in an Electric Field

In the case of a polar gas under the influence of an alternating field, the molecules are so far apart that viscosity effects should be negligible. Nevertheless, since an electric field exerts a torque on a permanent dipole, it seems logical that if the applied frequency were made to correspond to the rotational frequency of the molecule, there would result an interaction and consequent absorption of energy from the field. If the rotational frequency spectrum is considered to be continuous, as it is according to the classical viewpoint, these considerations indicate that there should be a smooth curve of absorption *versus* frequency (such as is observed for liquids). However, it is found experimentally that absorption occurs only at certain frequencies. Quantum theory has made it possible to explain this particular kind of absorption of electrical energy: in fact, a good example of the failure of classical and the success of quantum theory is provided by this problem. Accordingly, the quantum theory will be used in our discussion of polar gases: first, in consideration of the static dielectric constant, and then for the problem of absorption of energy from a time-varying field. A detailed discussion of quantum theory will not be attempted here; the emphasis will be on the conclusions to be drawn out of its applications.

Polar Gases in a Static Field

As the first step in our discussion of polar gases, we must present the wave mechanical equation. This may be obtained somewhat artificially through simple substitution for terms that occur in the expression for the classical Hamiltonian, discussed briefly in Chapter 2. The Hamiltonian, H, which for a conservative system represents total energy, is a function of Cartesian coordinates x, y, and z, and conjugate momenta p_x, p_y, p_z. For a single particle of mass m:

$$H = \frac{1}{2m}(p_x{}^2 + p_y{}^2 + p_z{}^2) + V(x, y, z)\,,$$

References p. 149

where V is potential energy. The 'prescription' for the quantum mechanical equation is to substitute for p_x (and similarly for p_y and p_z) the expression:

$$\frac{h}{2\pi i}\frac{\partial}{\partial x},$$

with the understanding that the resulting symbol $(\partial/\partial x)^2$ stands for the second derivative $\partial^2/\partial x^2$. The new 'Hamiltonian' function is then used as an operator applied to an unknown function ψ to yield the differential equation:

$$H\psi = U\psi$$

or:

$$-\frac{h^2}{8\pi^2 m}\left(\frac{\partial^2\psi}{\partial x^2}+\frac{\partial^2\psi}{\partial y^2}+\frac{\partial^2\psi}{\partial z^2}\right)+\psi V(x, y, z) = U\psi\,,$$

where U is the total energy of the system and V is potential energy. This is a form of the famous Schroedinger equation, the mechanical analogue of the optical wave equation.

The way in which this equation was developed is an interesting history. Because the optical wave equation formulated by Maxwell was able to explain the dispersion phenomena of light whereas geometrical optics was unable to do so, scientific thinkers early in this century wondered whether there was an analogous basic mechanical equation from which the classical mechanical laws could be obtained as a limiting case. The hope was to find a physical basis for the necessary but then foundationless assumption of the discreteness of energy in atomic systems. They were guided by the fact that the well known classical Hamilton-Jacobi equation for a single particle can be put into a form which exactly parallels the fundamental equation for geometrical optics. The idea was that if one could replace with mechanical parameters the optical ones such as frequency and phase velocity in the electromagnetic wave equation, one might obtain the desired basic mechanical wave equation.

The formulation of such a mechanical parameter corresponding to optical frequency seemed to be a permanent stumbling block; but then there was published by Louis de Broglie an illuminating paper dealing with the observed optical refraction of electrons[9]. He began by assuming that waves and particles are two different aspects of the same thing: a 'wave' associated with a moving particle should obey Fermat's optical principle at the same time that the particle itself obeys Hamilton's mechanical principle. In applying this idea to the description of electron diffraction, de Broglie formulated a relationship between the total energy, U, of the particle, and the frequency, ν, of the associated wave: $U = h\nu$, where h is Planck's constant. This is a generalization of Bohr's principle which relates the frequency of an optical phenomenon in an atomic system to the energy difference between the initial and final states.

Beginning with de Broglie's idea, Schroedinger [10] established the value of the phase velocity, v. By making use of the principle that the mechanical wave equation must reduce to the Hamilton-Jacobi equation in the limit of infinite frequency (or in the limit as h approaches zero), he obtained the relationship:

$$\frac{U^2}{v^2} = 2m(U - V),$$

where V is potential energy.

Upon introducing the values of v and ν into the optical equation, he obtained the fundamental equation for wave mechanics:

$$\text{div grad}\ \psi + \frac{8\pi^2 m}{h^2}(U - V)\,\psi = 0\,.$$

The function ψ represents the amplitude of some type of wave, and as in the optical situation, we say that the intensity of the wave at any point is given by the square of the amplitude, $|\psi|^2$. In dealing with the mechanical system, we shall accept the postulate that 'intensity' means probability; *i.e.*, if the function $|\psi|^2$ is normalized so that $\int_\tau |\psi|^2 \mathrm{d}\tau = 1$, where τ refers to all of space, then $|\psi|^2 \mathrm{d}\tau$ represents the probability that the particle is located in the volume element $\mathrm{d}\tau$. Accordingly, when $|\psi|^2$ is normalized, ψ is called the 'probability amplitude'.

Of all the solutions to Schroedinger's equation, the only physically meaningful ones are those for which ψ is finite and single-valued everywhere, and for only certain values of energy U will this be true. In speaking of these energies as a group, people have come to call them 'energy levels'. For systems of finite extent, the levels are always discrete (in contrast to the classical construction in which energy is continuous), although in many practical cases and always for macroscopic systems the group of levels includes so many different energies that the finest experimental measurements fail to detect the discontinuities.

The solution to Schroedinger's equation that we are interested in is the one that applies to a polar gas in an electric field. Let us consider a

References p. 149

molecule, like HCl, having a permanent moment μ; we may consider it (neglecting vibrations of the molecule) to be equivalent to two point masses connected by a rigid weightless rod of length r, having a permanent moment which points along the direction of the rod. The chlorine atom is so much heavier than the hydrogen atom that for all practical purposes we take the chlorine atom to be fixed at the origin of a set of coordinates. Because the hydrogen atom is thus constrained to move on the surface of a sphere, Schroedinger's equation is most conveniently expressed in polar coordinates:

$$\frac{1}{\sin \vartheta} \frac{\partial}{\partial \vartheta} \left(\sin \vartheta \frac{\partial \psi}{\partial \vartheta} \right) + \frac{1}{\sin^2 \vartheta} \frac{\partial^2 \psi}{\partial \varphi^2} + \frac{8\pi^2 I}{h^2} (U - V) \psi = 0 .$$

The original Laplacian operator (div grad) is represented in polar coordinates by the first two terms; the absence of a partial derivative with respect to r reflects the constraint that the distance between the two atoms be fixed. The equation has been multiplied through by the factor r^2 which appears in the denominator of the Laplacian when it is expressed in these coordinates; the resultant expression, mr^2, is the moment of intertia and has been designated as I. The advantage of this simple step is that the differential equation is thereby applied to the unit sphere: there is no explicit dependence on the original radius. If the potential energy V is due to an electrical field E, its value is $-\mu E \cos \vartheta$.

In the general case this equation can be solved only with great difficulty; but for the special case of vanishingly small fields, a comparatively simple method will produce a good approximation, provided that the solution is known for the case in which there is no external field at all. Accordingly, we must first find a solution for the case of no potential energy.

Fortunately, the differential equation that applies in the absence of potential energy is one which had been the subject of thorough mathematical investigation prior to the formulation of the particular problem we are considering. This equation, which we write as:

$$\frac{1}{\sin \vartheta} \frac{\partial}{\partial \vartheta} \left(\sin \vartheta \frac{\partial \psi}{\partial \vartheta} \right) + \frac{1}{\sin^2 \vartheta} \frac{\partial^2 \psi}{\partial \varphi^2} + n(n+1) \psi = 0 ,$$

was known to yield solutions that remain finite over the whole sphere surface only when n is some integer, including zero. This means that for our purposes the energy levels are immediately definable, for the product $n(n+1)$ can be identified with the term

$$\frac{8\pi^2 I}{h^2}\,U\,,$$

so that

$$U = \frac{n(n+1)\,h^2}{8\pi^2 I}\,.$$

In terms of a unit energy, ε, defined as $h^2/8\pi^2 I$, the only allowed levels, U_n, are thus given by:

$$U_n = n(n+1)\,\varepsilon\,.$$

Fig. 24 pictures the relationship of these levels. If we employ the technique

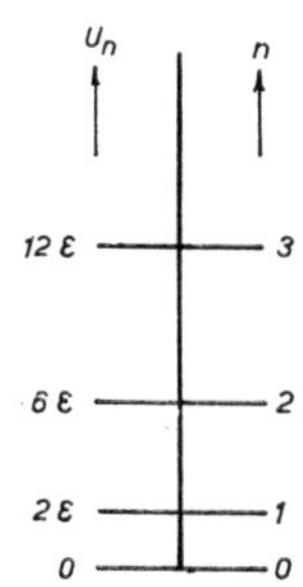

Fig. 24

of 'separation of variables', according to which we assume that ψ can be represented as the product $\Theta\,(\vartheta)\;\Phi\,(\varphi)$, then we are able to decompose the partial differential equation into two total differential equations. The one that involves $\Phi(\varphi)$ is readily solved by elementary methods and yields the two solutions, $\Phi = \cos\,(p\varphi)$, and $\Phi = \sin\,(p\varphi)$, where p is any whole number including zero. The other differential equation, in terms of $\Theta(\vartheta)$, also includes the parameter p and has for its solution the well known Associated Legendre Polynomials. These depend on both n and p and are generally abbreviated as $P_n{}^p\,(\vartheta)$ to show this dependence. Because of the boundary conditions that ψ must be single valued and finite, the integer p must be no greater than n: for a given value of n, the values of p may run from zero to n. Now let us consider simultaneously the informa-

References p. 149

tion that (a) the total energy depends only on n; (b) there are two different solutions to $\Phi(\varphi)$, namely $\cos(p\varphi)$ and $\sin(p\varphi)$; and (c) there are $n+1$ different solutions to $P_n{}^p(\vartheta)$ for a given value of n. The conclusion to be drawn is that there are, in all, $2n+1$ different functions belonging to each energy (excluding the trivial solution $\psi = 0$ arising out of $\sin(p\varphi)$ for $p = 0$). To describe this situation, one says that each ψ is $(2n+1)$-fold degenerate. As an illustrative example, we show the values of ψ belonging to the first two values of n, zero and unity:

$$n = 0 \qquad \psi = \quad P_0{}^0 = 1$$

$$n = 1 \qquad \psi = \begin{cases} P_1{}^0 & = \cos\vartheta \\ P_1{}^1 \cos\varphi & = \sin\vartheta\cos\varphi \\ P_1{}^1 \sin\varphi & = \sin\vartheta\sin\varphi \end{cases}$$

The last three expressions represent the projections of an arbitrary point on the sphere against the three coordinate axes x, y, and z. The name applied to these total solutions in terms of ϑ and φ is 'spherical harmonics'; a discussion of the meaning of spherical harmonics is given in the Appendix.

Probabilities associated with the various levels are obtained from the normalized ψ functions, called the probability amplitudes. To perform the normalization, we set: $\int |\psi|^2 \,\mathrm{d}\Omega = 1$. Thus, for the case of zero potential energy (no applied field), we find that for $n = 0$ the probability amplitude is: $\psi = 1/\sqrt{4\pi}$; for $n = 1$, the normalizing factor is $\sqrt{3/4\pi}$. The probability that the dipole will be found within a particular solid angle $\mathrm{d}\Omega$ is given by $|\psi|^2 \,\mathrm{d}\Omega$. This reduces to $(1/4\pi)\,\mathrm{d}\Omega$ for $n = 0$, indicating that the probability depends only on the size of $\mathrm{d}\Omega$, or, equivalently, that there is no preferred orientation in the ground state. For the first level, the total $|\psi|^2$ is made up of three sums, one for each degeneracy:

$$\frac{3}{4\pi}\left[\cos^2\vartheta + \sin^2\vartheta\cos^2\varphi + \sin^2\vartheta\sin^2\varphi\right].$$

This quantity is constant, which means that in this level, too, no preferential dipole orientation exists. The same result is found for any level n. This is as expected since in the absence of a field there is nothing to influence dipole direction.

The next step in our problem is to solve the Schroedinger equation when the external field is present. The presence of the field gives rise to an angular-dependent potential, $-\mu E \cos \vartheta$, as discussed above. If we limit ourselves to the consideration of very small external fields, we may use the method of successive approximations to calculate the new probability amplitude ψ, which is thus assumed to be the sum of the amplitude at zero potential energy, plus a small correction term. Details of this calculation are presented in *Polar Molecules* [11], where it is shown that the correction term is of the order E^2, for a diatomic molecule. In terms of energies, the results are that if the energy U and the magnitude of the potential energy are both expressed in terms of the unit energy ε, *i.e.*, $U = \lambda\varepsilon$ and $\mu E/\varepsilon = \kappa$, where λ and κ are numerical parameters, then for the ground state:

$$\lambda_0 = -\kappa^2/6\,,$$

and for the nth level:

$$\lambda_n = n(n+1) + \frac{\kappa^2}{2}\left[\varphi_n{}^p + \varphi_{n+1}{}^p\right],$$

where

$$\varphi_n{}^p = \frac{(n-p)(n+p)}{n(2n-1)(2n+1)}\,.$$

Each level (which was $2n+1$ degenerate in the absence of a field) is now split into a group of levels clustered closely around the original value; each of these is at most doubly degenerate. It turns out that the 'center of gravity' of energy for all levels but the lowest remains unchanged; the ground level energy is lowered by an amount

$$-\frac{\kappa^2}{6}\varepsilon\,.$$

This situation is illustrated in Fig. 25 for $n = 0$ and $n = 1$. This means, as we shall show, that each group of levels (except for the ground level) can be treated in a good first approximation as a single level having an energy equal to the average: thus the levels are still considered to have degeneracies of $2n + 1$, and their energies are considered to be the same as in the absence of a field.

These energies and degeneracies are required for the calculation of the population (or probability) of the various levels (this is discussed to some extent in Chapter 2). The population of any level is given by the Boltzmann formula, with the requirement that each energy exponential must be multiplied by the corresponding degeneracy. Thus the population of each level (other than the ground level) is proportional to the quantity:

$$(2n + 1)\,e^{-U_n/kT} = (2\,n + 1)\,e^{-n(n+1)\,\varepsilon/kT}\,.$$

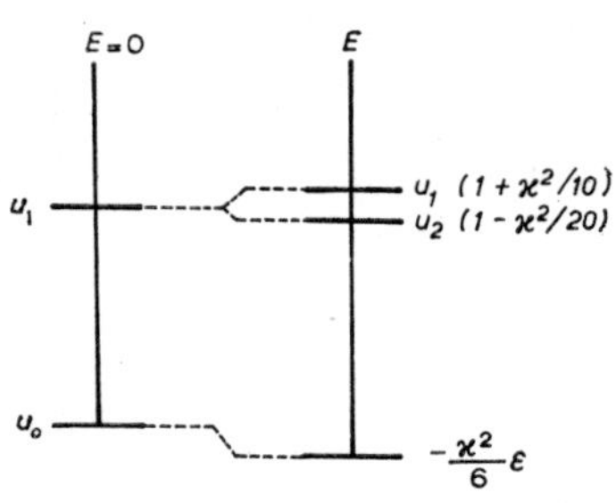

Fig. 25

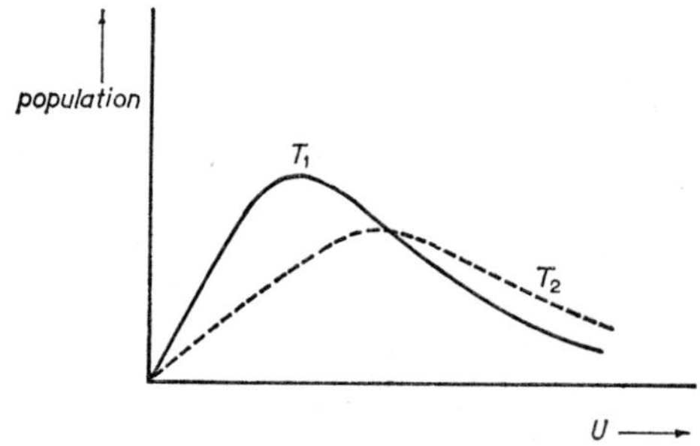

Fig. 26

A change in population with temperature is indicated by the formula, and this is shown qualitatively in graphical form in Fig. 26, where for simplicity a smooth curve is drawn through the points representing distinct energies. The population reaches a maximum at a level which represents the compromise between increasing probability arising from the degeneracy factor $2n + 1$, and decreasing probability resulting from the increasingly negative exponential.

The polarization of a polar gas may be calculated from knowledge of how the populations of the energy levels change upon application of an electric field. Thermodynamic reasoning combined with the evaluation of parameters through statistical methods yields the answer in a straightforward manner. The details can be found in the text, *Polar Molecules* [12]; a brief summary will be given here.

The combined first and second laws of thermodynamics read:

$$dU = \delta Q + \delta W = TdS + \delta W\,,$$

where U is total energy, Q is heat, and W represents all the energy involved other than heat. In this case, δW is the sum of the change in potential energy plus the work done by the field in changing the internal energy of the gas. The change in potential energy arising out of a variation in the field strength is given, on a unit volume basis, by $-\mathrm{d}(PE)$, since the total polarization P depends on field strength. The 'internal' work accomplished by the change of field strength is obtained from the expression embodying the Faraday point of view of how energy is stored in an electrostatic field, as discussed in Chapter 1. It was shown that $E\mathrm{d}D/4\pi$ is the energy increment per unit volume of dielectric; E represents the applied field strength and D is the electrical displacement. Even in the complete absence of dielectric, there is a change of energy per unit volume, of magnitude $E\mathrm{d}E/4\pi$; thus the difference

$$\frac{E\mathrm{d}D}{4\pi} - \frac{E\mathrm{d}E}{4\pi}$$

represents the 'internal' work expended on the dielectric. This is more conveniently written

$$\frac{E\mathrm{d}(D-E)}{4\pi},$$

and from the general equation relating displacement, field strength, and polarization per unit volume, namely $D = E + 4\pi P$, one finds that $E\mathrm{d}P$ is the 'internal' work per unit volume of dielectric. The entire change in work, δW, being the sum of internal and potential energy terms, can be written:

$$\delta W = E\mathrm{d}P - \mathrm{d}(PE) = -P\mathrm{d}E\,.$$

Now we also know from thermodynamics that we can define a thermodynamic potential Φ in the manner $\Phi = U - TS$. (This function Φ is sometimes designated as free energy, F; we used it in this way in Chapter 2.) At constant temperature, $\mathrm{d}\Phi$ represents the increment of total work done on the system*:

$$\mathrm{d}\Phi = \delta W = -P\mathrm{d}E\,,$$

or:

$$P = -(\partial\Phi/\partial E)_T\,.$$

* The mathematical manipulation is as follows. We write: $\mathrm{d}\Phi = \mathrm{d}U - T\mathrm{d}S - S\mathrm{d}T$. Then, because $\mathrm{d}U = T\mathrm{d}S + \delta W$, we write: $\mathrm{d}\Phi = T\mathrm{d}S + \delta W - T\mathrm{d}S - S\mathrm{d}T$. At constant T, $\mathrm{d}\Phi = \delta W$.

The quantity P, which is volume polarization of the dielectric, is the desired parameter.

The polarization P could be calculated without bringing into the discussion the notion of thermodynamic potential, by evaluating the terms in the formula:

$$\delta W = \mathrm{d}U - T\mathrm{d}S = -P\mathrm{d}E\,.$$

Variations of both the average total energy and the entropy are intimately connected with and completely derivable from the energy level scheme. The advantage of dealing with the potential Φ lies in its very simple representation as $\Phi = -kT \ln \sum$, where $\sum$ is the partition function. If we are considering a very dilute gas, in which it is assumed that there are no interactions between molecules, we can represent the system as an individual molecule*: partition function is**

$$\sum_{i=0}^{\infty} \mathrm{e}^{-U_i/kT}\,,$$

and the quantity $-(\partial\Phi/\partial E)_T$ is the expression for average moment for a molecule, m, rather than for total polarization per unit volume, P. The energies U_i are functions of the field strength E, and so Φ is expressed in the proper variables for the differentiation.

If the energy levels obtained from Schroedinger's equation are introduced, the partition function must be written so as to include degeneracy. The contribution of the ground state is $\mathrm{e}^{-U_0/kT}$; in the presence of a field U_0 is

$$-\frac{\kappa^2}{6}\varepsilon \text{ or } -\frac{\mu^2 E^2}{6\varepsilon}\,.$$

At the higher energy levels, though, the field has, to excellent approximation, no influence on the partition function. If we consider, for example, the energy level U_1, which is threefold degenerate, we find that its

* The correct answers will be obtained only if the number of molecules remains fixed; otherwise, we would have to cope with Gibb's paradox, discussed in Chapter 2.

** The notation in this expression for partition function is slightly different from that used in Chapter 2. There the energies were written as u_i rather than U_i.

contribution to the partition function is $3e^{-U_1/kT}$ in the absence of a field. When a field is applied, the new contribution is:

$$2e^{-U_1\,(1-\kappa^2/20)/kT} + e^{-U_1\,(1+\kappa^2/10)/kT}\,,$$

since the level splits to a twofold and a single level. This is:

$$e^{-U_1\,/kT}\left[2e^{\kappa^2 U_1\,/20kT} + e^{-\kappa^2 U_1\,/10kT}\right].$$

Expansion of the exponential yields:

$$3e^{-U_1\,/kT}\left[1 + \ldots\right],$$

where the second term in the bracket is of the order κ^4 and may be dropped. In other words, we obtain again, to a good approximation, the expression that applies in the absence of a field. Thus total partition function, expressed as the sum of contributions from the ground state and from the higher energy states, is:

$$\sum = e^{\mu^2 E^2/6\varepsilon kT} + \sum_{n=1}^{\infty} (2n+1)\, e^{-n\,(n+1)\,\varepsilon/kT}\,.$$

Upon performing the differentiation indicated by the relation $m = -(\partial\Phi/\partial E)_T$, we find that the average moment for a molecule is given by the expression:

$$m = \frac{\dfrac{\mu^2}{3\varepsilon}E}{\sum\limits_{n=0}^{\infty} (2n+1)\, e^{-n\,(n+1)\,\varepsilon/kT}}\,.$$

The conclusion to be drawn is that the influence of a field on the value of moment is exerted through the energy level shift of the ground state only.

In connection with the problem of calculating the average moment, it is enlightening to investigate what orientation is given by the field to molecules in the various levels. The ψ function for the lowest state may be calculated by a 'perturbation method' involving the lowest power non-vanishing term in E. It turns out that the wave function for the ground state is proportional to

$$1 + \frac{1}{2}\frac{\mu E}{\varepsilon}\cos\vartheta\,;$$

References p. 149

i.e., the probability for a particular direction, $|\psi|^2 \, d\Omega$, is proportional to

$$\left[1 + \frac{\mu E}{\varepsilon} \cos \vartheta\right] d\Omega ,$$

and the molecule has a preferential orientation in the field direction. However, one finds for the higher energy levels that because the 'center of gravity' of each group of split levels is unchanged, there is no preferential orientation occurring in these higher levels. This result is really not surprising if one cares to interpret the quantum mechanical picture with the rather unsafe aid of a mechanical model: a physical interpretation is that only those molecules that are 'standing still' react with the external field.

A comparison between the evaluations of moment as derived from classical and from quantum considerations may now be made. This may be done with greatest clarity if the quantum expression is put into the form:

$$m = \frac{\mu^2}{3kT} E \frac{1}{x \sum\limits_{n=0}^{\infty} (2n + 1) \, e^{-n(n+1)x}} ,$$

where x stands for ε/kT. The fundamental energy ε is very small compared with kT: even for HC1, which has a small moment of inertia and hence a large ε, this ratio is $14.7/T$. The sum can therefore be approximated by an integral in the following manner:

$$x \sum_{n=0}^{\infty} (2n + 1) \, e^{-n(n+1)x} \, \Delta n \approx x \int_0^{\infty} (2n + 1) \, e^{-n(n+1)x} \, dn ,$$

where Δn is unity in the summation. Designating $n(n + 1)$ as φ, we can write that $d\varphi = (2n + 1) \, dn$, and the integral transforms to the elementary form:

$$x \int_0^{\infty} e^{-x\varphi} \, d\varphi = x \cdot \frac{1}{x} = 1 .$$

We conclude, then, that in the limit of very small ε (which would correspond to the approach to zero of Planck's constant, h), the formula for m reduces to:

$$m = \frac{\mu^2 E}{3kT} .$$

This is identical to the classical result. A second approximation, in which the finite value of ε is considered, leads to the expression:

$$m = \frac{\mu^2 E}{3kT}\left[1 - \frac{1}{3}\frac{\theta}{T}\right],$$

where θ stands for ε/k and represents a characteristic temperature pertaining to the molecule (it is 14.7° K for HCl). We conclude, then, that the classical calculation is fundamentally wrong except for high temperatures; for most molecules, however, ordinary room temperatures are sufficiently high to make negligible the term

$$\frac{1}{3}\frac{\theta}{T}.$$

In conclusion, we can make the following observations. The value of the static dielectric constant of a diatomic polar gas is determined by the behavior of the molecules in the ground state. The actual energy value associated with this state depends on the square of the applied field. The observed decrease in dielectric constant with rising temperature is explained by the increased population of the higher levels at the expense of the all-important lower one. For sufficiently high temperatures the population of the lowest level, and thus the dielectric constant of the gas of permanent dipoles, is well approximated by the classical inverse temperature law.

Polar Gases in an Alternating Field

The description of the behavior of a polar gas under the influence of an alternating field requires more general quantum mechanical considerations than we employed before. For static conditions, the problem was mainly the evaluation of energy levels. For the case of an alternating field, we must also consider the probability amplitude ψ and the quantum mechanical frequency ν that is associated with the level energy U, according to the expression $\nu = U/h$.

A dipolar gas under the influence of an alternating field will interact strongly with the field at certain frequencies: there will be emission or absorption of energy. We can investigate this phenomenon by formulating an expression for the 'transition probability', *i.e.*, the probability that a

References p. 149

molecule under the influence of radiation can be made to 'jump' from a given level to another one, higher or lower than the first. Because the description of a state of the system is contained in the properties of its corresponding wave function, Ψ, which is $\psi\, e^{i\omega t}$ or $\psi\, e^{i2\pi(U/h)t}$, transition probabilities should be derivable through quantum mechanical procedure.

Now the extent of interaction with radiation of a given frequency is proportional to the square of the dipole moment oscillating at that frequency†, and so to calculate transition probability we need to calculate the magnitude of this dipole moment. This can be calculated from the time-varying part of the electric charge distribution over the surface of the sphere described by the rotating dipole. It has been postulated that probability density for electric charge is proportional to $|\Psi|^2$ or, equivalently, to $\Psi\Psi^*$, where the asterisk refers to the complex conjugate. If we suppose that the molecule is undergoing a transition, the probability amplitude should be a superposition of the individual amplitudes:

$$\begin{aligned}\Psi &= C_{n',p'}\,\Psi_{n',p'} + C_{n,p}\,\Psi_{n,p} \\ &= C_{n',p'}\,\psi_{n',p'}\,e^{i2\pi(U_{n'}/h)t} + C_{n,p}\,\psi_{n,p}\,e^{i2\pi(U_n/h)t}\,.\end{aligned}$$

Here the primes refer to the final state; the C's are the corresponding coefficients. Thus the quantity $\Psi\Psi^*$ is:

$$\begin{aligned}\Psi\Psi^* &= C_{n',p'}{}^2\,\psi_{n',p'}{}^2 + C_{n,p}{}^2\,\psi_{n,p}{}^2 \\ &\quad + 2C_{n',p'}\,C_{n,p}\,\psi_{n',p'}\,\psi_{n,p}\cos\frac{2\pi}{h}(U_{n'} - U_n)t\,,\end{aligned}$$

where the final term represents the time-varying part of the charge distribution. The frequency of the radiation involved in the transition is given by $(U_{n'} - U_n)/h$. The magnitude of the dipole moment involved is obtained by averaging dipole moment over all orientations:

$$m = 2C_{n',p'}\,C_{n,p}\int_{\Omega}\psi_{n',p'}\,\psi_{n,p}\,\mu\cos\vartheta\,d\Omega\,,$$

where μ is the permanent moment of the molecule, and ϑ is the angle measured from the field direction. The integral in this equation must

† The calculation made according to the Maxwell equations for a vibrating dipole (given in the Appendix of Chapter 4) shows that the intensity of radiation emitted by the dipole is proportional to the square of its vibrating dipole moment.

have a non-zero value if interaction with radiation is to occur. The square of this integral is defined as the transition probability. Since we are generally interested in the limiting case of small external fields, the unperturbed ψ functions can be used in the calculations.

However, in the search for possible transitions, the possibilities may be quickly narrowed down even before a single transition probability is calculated, by application of well known properties of spherical harmonics: details may be found in *Polar Molecules*[13]. One finds that unless certain relationships hold between the initial and final values of the parameters n and p (which identify the ψ functions), orthogonality properties cause the integrals to yield zero identically. From these relationships are derived the so-called 'selection rules'. Transitions will occur only when $n' = n \pm 1$ and $p' = p$ or $p \pm 1$; for any other pairs of n's and p's, energy will not be absorbed or emitted, and the term 'forbidden transition' applies. The energy difference between the two states depends only upon n, and so the transition frequency is simply

$$\left[n(n+1) - n(n-1)\right]\frac{\varepsilon}{h},$$

or

$$\frac{2n\varepsilon}{h},$$

or

$$2n\,\frac{h}{8\pi^2 I}.$$

Thus the position of absorption lines is established: they will occur at

$$\frac{2\varepsilon}{h}, \frac{4\varepsilon}{h}, \frac{6\varepsilon}{h},$$

etc.

The polarizability of a polar gas in an alternating field can also be calculated through wave mechanical considerations. For this case, in which the potential energy of the gas molecule in the field is time dependent, we cannot employ the Schroedinger wave equation as it stands, but must generalize it to include time variation. We can put the equation into the proper form by expressing potential energy V as $-\mu E_0 \cos \vartheta\, e^{i2\pi\nu t}$, where ν is the frequency of the applied field, and by using an alternate way of expressing the total energy U in accordance with the relation

$$\frac{\partial \Psi}{\partial t} = \frac{2\pi \mathrm{i}}{h} U \Psi .$$

Thus the wave equation assumes the form:

$$\text{div grad } \Psi + 2I \frac{2\pi \mathrm{i}}{h} \frac{\partial \Psi}{\partial t} - 2I \left(\frac{2\pi \mathrm{i}}{h} \right)^2 \mu E \cos \vartheta \Psi = 0 ,$$

where I represents the moment of inertia and the operation div grad is to be performed on the sphere of unit radius; the symbol for the applied field, E, is an abbreviation for $E_0 \, \mathrm{e}^{\mathrm{i} 2\pi \nu t}$. For small values of applied field strength, we assume that Ψ can be represented as the sum $\Psi_0 + G$, where Ψ_0 satisfies the above equation in the absence of the field, and G is a small correction. The result obtained in this manner is given in detail in *Polar Molecules* [14].

Having calculated Ψ, we would then be in a position to calculate the value of the moment vibrating in the same direction and with the same frequency as that of the applied field. The average value of this moment is:

$$m(n, p) = \int_{\Omega} \Psi_{n,p} \Psi^*{}_{n,p} \, \mu \cos \vartheta \, \mathrm{d}\Omega = \int_{\Omega} | \psi_{n,p} |^2 \mu \cos \vartheta \, \mathrm{d}\Omega \, . \, \dagger$$

This gives the desired moment as a function of the parameters p and n. What we really want is the average value of this moment taken over all the energy levels, and this is obtained by the usual statistical method based on the Boltzmann distribution. The moment arising out of any level characterized by n and p is proportional to $m(n, p) \, \mathrm{e}^{-n(n+1)\varepsilon/kT}$, and the average, $\overline{m}$, is obtained by summing this quantity, first over all the values of p for fixed n, and then over all n.

The final expression gives the average polarizability α (average moment divided by field strength) as a function of frequency ν:

$$\alpha = \frac{\mu^2}{3kT} \sum_{n=1}^{\infty} \frac{1}{1 - (\nu/2n\nu_0)^2} \left[\mathrm{e}^{-n(n-1)\frac{\theta}{T}} - \mathrm{e}^{-n(n+1)\frac{\theta}{T}} \right] .$$

† It is to be noted that the integration of $\int_{\Omega} | \psi_{n,\,p} |^2 \mu \cos \vartheta \, \mathrm{d}\, \Omega$, if performed using only the unperturbed ψ functions, yields zero identically. To obtain a finite value for $m(n, p)$, it is necessary to take into account the correction term G. In contrast, it was not necessary to do this for the calculation of transition probability because in that calculation wave functions associated with different states were combined.

Here the frequency ν_0 is defined according to the relation $h\nu_0 = \varepsilon$, where ε is the energy constant for the molecule: for the case of HCl, the frequency ν_0 is calculated to be 3×10^{11} sec^{-1}. The characteristic temperature θ is in like manner defined by the relation $k\theta = \varepsilon$, and for HCl is calculated to be 14.7° K. A more exact formula would contain the factor

$$\left(1 - \frac{1}{3}\frac{\theta}{T} + \ldots\right),$$

the familiar correction to the integral approximation of the sum representing the partition function in the quantum mechanical scheme of discrete energy.

We see from the expression for α that the polarization approaches infinity when the frequency ν corresponds to the transition frequencies $2n\nu_0$, which are $2\nu_0$, $4\nu_0$, $6\nu_0$, etc. Thus the system behaves as though it consisted of resonators with natural frequencies $2n\nu_0$.

Some Examples of the Application of Dispersion Theories

The contrast between the behavior of dipolar gases, and that of polar molecules in viscous liquid media is illustrated in Figs. 27 and 28. Regions in which dielectric constant rises with increasing frequency are referred to as regions of normal dispersion; where dielectric constant falls as frequency increases, there is anomalous dispersion. The dispersion characteristics of the gaseous state can be attributed to resonance as described through quantum mechanics; for the liquid state, classical

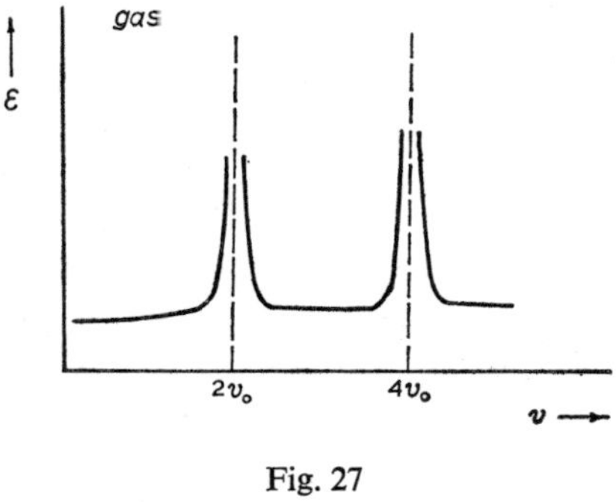

Fig. 27

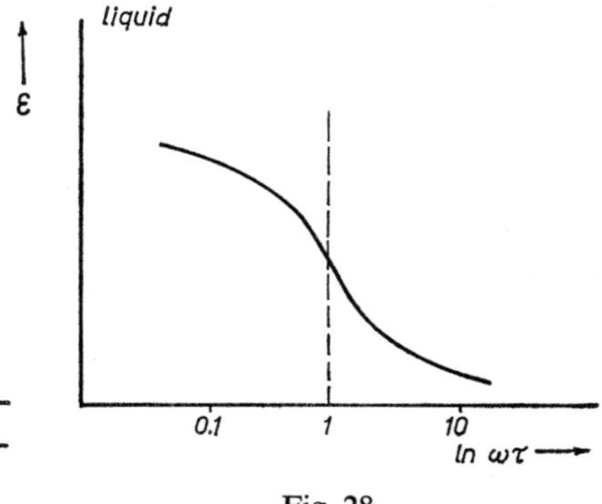

Fig. 28

References p. 149

considerations are adequate to describe anomalous dispersion through the relaxation effect.

There is another interpretation of frequency dependence of dielectric constant that is worthy of note. This is Cauchy's explanation of dispersion (set forth in the 1830's), based on classical ideas of forced mechanical vibration of a molecule under the influence of electromagnetic radiation. The development of Cauchy's theory may be presented as follows. Let us consider a mass m which is carrying a charge e and which is connected to one end of a spring having a force constant f, the other end of the spring being fixed. Suppose a harmonic electrical force of magnitude $E\mathrm{e}^{\mathrm{i}\omega t}$ is applied to the mass in a direction parallel to the spring axis. If x is the displacement of the mass from equilibrium, the equation of motion is:

$$\ddot{x} + \omega_0^2 \, x = \frac{e}{m} \, E \, \mathrm{e}^{\mathrm{i}\omega t} \, ,$$

where $\omega_0 = \sqrt{f/m}$; ω_0 is the proper frequency of vibration for the system. We try the solution $x = A\mathrm{e}^{\mathrm{i}\omega t}$, where A is assumed to be time independent, and find that

$$A = \frac{(e/m) \, E}{\omega_0^2 - \omega^2} \, .$$

The electric moment is ex, or

$$\frac{(e^2/m) \, E \, \mathrm{e}^{\mathrm{i}\omega t}}{\omega_0^2 - \omega^2} \, ;$$

and the polarizability is then

$$\frac{e^2/m}{\omega_0^2 - \omega^2} \, .$$

Using the classical expression for the dielectric constant of a gas, $\varepsilon - 1 = 4\pi n\alpha$, and with a generalization to include harmonic oscillators of various proper frequencies, we obtain:

$$\varepsilon - 1 = r^2 - 1 = 4\pi \sum n \, \frac{e^2/m}{\omega_0^2 - \omega^2} \, ,$$

where r is refractive index.

A more realistic expression would include the effect of friction, as:

$$\ddot{x} + p\dot{x} + \omega_0^2 x = (e/m)\, E\, \mathrm{e}^{\mathrm{i}\omega t}\,;$$

the frictional force depends on the velocity, and p is the proportionality constant. We try as a solution $x = B\,\mathrm{e}^{\mathrm{i}\omega t}$, since the equation is linear. After solving for B and substituting, we find:

$$x = \frac{(e/m)\, E}{\left[(\omega_0^2 - \omega^2)^2 + \omega^2 p^2\right]^{1/2}}\, \mathrm{e}^{\mathrm{i}(\omega t - \varphi)}\,,$$

where φ is defined according to the relation: $\tan\varphi = \omega p/(\omega_0^2 - \omega^2)$ *.

Fig. 29, which is a graph for the dispersion of glass, illustrates the

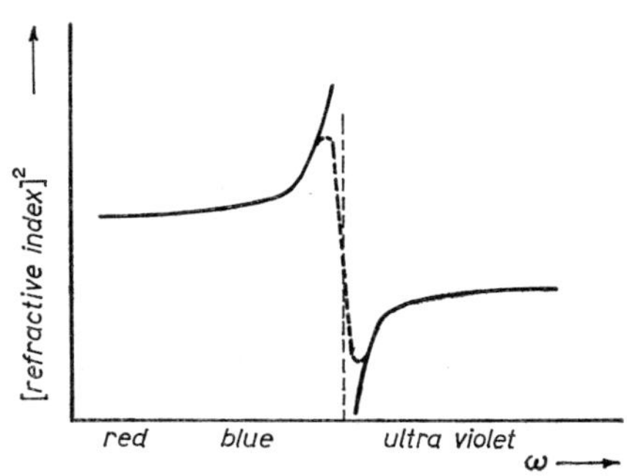

Fig. 29

theory. The solid curve, which represents the case for the absence of friction, predicts infinite dielectric constant at the proper frequency. The inclusion of an effect due to friction results in the prediction of a region of anomalous dispersion, shown by the dotted curve. This diagram fits the actual dispersion characteristics of glass quite well.

For the region of frequency where $\omega^2/\omega_0^2 \ll 1$, the visible region in this case, the effect of the friction term is negligible and the expression is easily developed into the form:

$$r^2 - 1 = 4\pi n \frac{e^2}{m\omega_0^2}\left[1 + \left(\frac{\omega}{\omega_0}\right)^2\right].$$

* This follows from the identity $A + \mathrm{i}B = (A^2 + B^2)^{1/2}\, \mathrm{e}^{\mathrm{i}\varphi}$, where φ is the angle defined as $\tan\varphi = B/A$.

References p. 149

As an example of the application of this formula, we shall calculate the refractive index of water at near ultraviolet frequencies. (In this region of the spectrum, the frequency is so great that only the electrons can follow the field and contribute to the refractive index.) The observed value of proper frequency, in terms of cycles per second, $\omega_0/2\pi$, is 3×10^{15}, which is in the ultraviolet region, and so we shall have $\omega^2/\omega_0^2 \ll 1$. Taking for liquid water $m = 9\times10^{-28}$ gm, and $n = \frac{1}{3}\times10^{23}$ cm^{-3}, and taking $e = 4.8\times10^{-10}$ esu, we find (neglecting interactions) that

$$r^2 = 1 + 4\pi n \frac{e^2}{m\omega_0^2} \approx 1\frac{1}{3}.$$

Hence refractive index, r, is about $1\frac{1}{6}$, and although this is in only fair agreement with the observed value of 1.33, it shows that such a simple model does give a reasonable result.

The determination of proper frequency can be accomplished by the simple and rather spectacular method of 'rest rays' developed by Rubens. Suppose that there are n sodium chloride crystals arranged in two rows in such a way (as sketched in Fig. 30) that light can be reflected from one

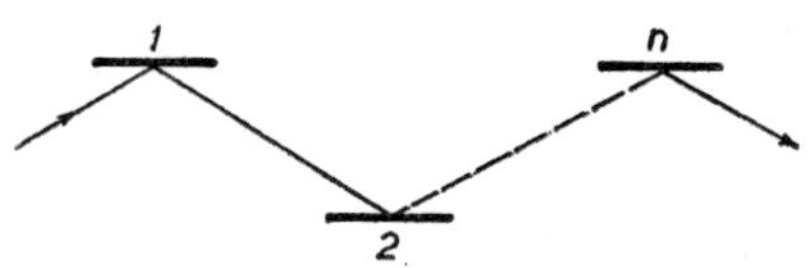

Fig. 30

surface to another. The amount of light reflected from each surface depends on the magnitude of the refractive index of the crystal, and refractive index rises sharply near the proper frequency (the more sharply, the less friction there is present). Because the amount of light finally reflected from the nth surface is proportional to the individual surface reflectivity raised to the nth power, the intensity of the outgoing ray rises very sharply when the frequency of the incoming beam approaches the proper frequency. This is the basis of the method of 'rest rays'.

(i) *The Boundary Condition for the Electrical Displacement D in the Absence of Free Charge*

Suppose that there are two adjacent regions having different dielectric constants, as illustrated in Fig. 31. Let us describe a rectangular cylinder

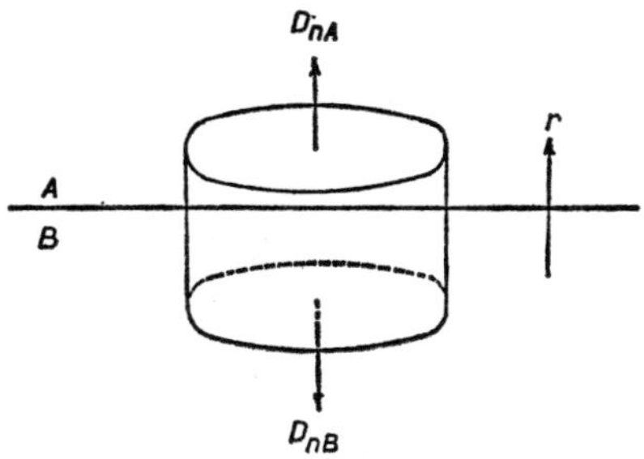

Fig. 31

in the shape of a pillbox with one face in region A and one in region B, and of negligible height. As was shown in Chapter 1, the displacement D is related to charge Q according to the expression: $\int_\sigma D_n \, d\sigma = 4\pi Q$; the integration is performed over the surface σ enclosing the volume, D_n refers to the component of D directed along the outward normal to the surface, and Q is the total free charge contained within the volume. When there is no free charge in the dielectric, $Q = 0$ and so $\int_\sigma D_n \, d\sigma = 0$. Suppose the area of each face is a; then, because the height is negligible, the integral may be written in the simple form: $(D_{nA})a + (D_{nB})a = 0$, or $D_{nA} = - D_{nB}$. Referred to a common direction, r, the relationship becomes:

$$D_{rA} = D_{rB} \,.$$

The conclusion is that in the absence of free charge, the normal components of displacement on two sides of a surface are equal.

References p. 149

(*ii*) *The Connection between Electrical Loss of Energy and the Imaginary Part of the Dielectric Constant*

From the ideas developed in Chapter 1, we know that the element of work, $\mathrm{d}W$, involved in a change of displacement from D to $D + \mathrm{d}D$ under the influence of a field is:

$$\mathrm{d}W = \frac{E\,\mathrm{d}D}{4\pi}\,.$$

We also know that D is εE; that $\varepsilon = \varepsilon_1 - \mathrm{i}\varepsilon_2$; and that E is $E_0 \mathrm{e}^{\mathrm{i}\omega t}$. If we take the real part of the product εE (since this represents the physical quantity), we find:

$$D = \varepsilon_1\, E_0 \cos(\omega t) + \varepsilon_2\, E_0 \sin(\omega t)\,.$$

Then we obtain for the instantaneous element of work involved:

$$\mathrm{d}W = \left[-\frac{\varepsilon_1\, E_0{}^2}{4\pi} \sin(\omega t)\cos(\omega t) + \frac{\varepsilon_2\, E_0{}^2}{4\pi}\cos^2(\omega t)\right]\omega\,\mathrm{d}t\,.$$

The first term contains the factor $\frac{1}{2}\sin(2\omega t)$, which, averaged over one cycle, yields zero; this term represents a periodic transfer of energy from the external source to the field and back again. However, the average of the term $\cos^2(\omega t)$, present as a factor multiplying ε_2, is not zero, but one half. This term must represent the actual dissipation of energy through friction. The energy loss per unit time is, then,

$$\frac{\omega\varepsilon_2\, E_0{}^2}{4\pi}\cos^2(\omega t)\,.$$

The electric current necessary to produce an equivalent power loss under the application of the field $E = E_0 \cos(\omega t)$ is, then:

$$I = \frac{\omega\varepsilon_2\, E_0}{4\pi}\cos(\omega t)\,.$$

This represents the actual 'loss current' that flows in the external circuit; *i.e.*, the current that flows in phase with the applied field. Also flowing in the external circuit is a current of magnitude

$$\frac{\omega \varepsilon_1 E_0}{4\pi}$$

that leads the applied voltage by a phase angle $\pi/2$; this purely capacitive current produces no loss of power.

(iii) Some Insight concerning the Meaning of Spherical Harmonics

There is an interesting physical meaning connected with spherical harmonics, which can be shown through the development of certain solutions to Laplace's equation.

It is known that if any solution to Laplace's equation has been found, others may be obtained from it by differentiation. For example, suppose that Φ is such a solution, *i.e.*, that $\Delta\Phi = 0$; then it is true that

$$\frac{\partial}{\partial x}\Delta\Phi = \Delta\frac{\partial\Phi}{\partial x} = 0\,.$$

Hence $\partial\Phi/\partial x$ is also a solution to Laplace's equation; and the partial derivatives in the other two directions are likewise solutions. This differentiation process may be repeated as often as desired.

Now let us apply this reasoning to an expression for potential in terms of distance, r; the expression is $\Phi = 1/r$, which is a solution to Laplace's equation for a unit charge located at the origin of coordinates. The potential we obtain by partial differentiation of the function $\Phi = 1/r$ with respect to x is

$$\frac{\partial\Phi}{\partial x} = -\frac{\cos\vartheta}{r^2}\,,$$

where $\cos\vartheta$ is x/r. As discussed in Chapter 1, this is the expression for the long-distance potential of a dipole of unit strength pointing in the negative x direction. Similarly, $\partial\Phi/\partial y$ and $\partial\Phi/\partial z$ represent unit dipoles pointing in the negative y and z directions. In other words, the system of two opposite charges separated by a small distance gives rise to a long-range potential which is described by the first partial derivatives of the

References p. 149

single charge potential. Likewise, it turns out that the second partial derivatives represent potentials arising from systems made up of two oppositely directed dipoles separated by a small distance; these are called 'quadrupoles', and exist in two types, the linear and the rectangular (see Fig. 32)*. Third derivatives represent octupoles (combinations of oppositely directed quadrupoles), and so on.

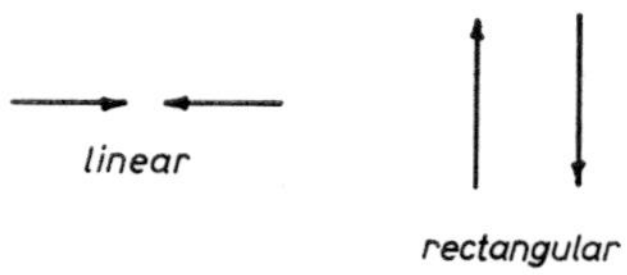

Fig. 32

All these results may be summarized in a single expression. We construct a general potential by adding together all the individual potentials, each multiplied by an arbitrary constant:

$$\Phi_T = \Phi + a_1 \frac{\partial \Phi}{\partial x} + a_2 \frac{\partial \Phi}{\partial y} + a_3 \frac{\partial \Phi}{\partial z} + \frac{1}{2}\left[b_{11} \frac{\partial^2 \Phi}{\partial x^2} + b_{22} \frac{\partial^2 \Phi}{\partial y^2} \right.$$

$$\left. + b_{33} \frac{\partial^2 \Phi}{\partial z^2} + 2b_{12} \frac{\partial^2 \Phi}{\partial x \partial y} + 2b_{13} \frac{\partial^2 \Phi}{\partial x \partial z} + 2b_{23} \frac{\partial^2 \Phi}{\partial y \partial z} \right] + \ldots,$$

where $\Phi = a_0/r$. If this is applied to a unit sphere, where $r = 1$, the values of the nth derivatives do not include a factor r^{-n-1}. And if the charge is a unit charge, so that a_0 is also unity, the derivatives are identifiable as the spherical harmonics of order n as first introduced by Maxwell; they are identical to the coefficients of the solution (in powers of $1/r$) of the differential equation presenting spherical harmonics.

Instead of beginning with a system of multipoles at the origin, let us now begin with a point charge e located at the coordinates ξ, η, and ζ. The potential at a point P, which is located by the coordinates x, y, and z, is:

$$\Phi(P) = e/\sqrt{(x-\xi)^2 + (y-\eta)^2 + (z-\zeta)^2}\,.$$

* Because a difference requires one term to be of the opposite sign from the other, the physical interpretation of the results obtained by differentiating the original potential for a single pole is that there must be an accumulation of identical but oppositely directed structures: two 'oppositely directed' charges give a dipole, two oppositely directed dipoles give a quadrupole, etc.

If, however, we wish to express this potential in terms of the distance r from the point P not to ξ, η, and ζ, but rather to some other position, we may expand $\Phi(P)$ in a Taylor series around that position. Accordingly, we designate this new reference point as the origin, and let Φ_0 represent the potential evaluated there. The expression for potential becomes:

$$\Phi(P) = \Phi_0 + \frac{\partial \Phi}{\partial \xi}\xi + \frac{\partial \Phi}{\partial \eta}\eta + \frac{\partial \Phi}{\partial \zeta}\zeta + \frac{1}{2}\left[\frac{\partial^2 \Phi}{\partial \xi^2}\xi^2 + \frac{\partial^2 \Phi}{\partial \eta^2}\eta^2 + \frac{\partial^2 \Phi}{\partial \zeta^2}\zeta^2 + 2\frac{\partial^2 \Phi}{\partial \xi \partial \eta}\xi\eta + 2\frac{\partial^2 \Phi}{\partial \xi \partial \zeta}\xi\zeta + 2\frac{\partial^2 \Phi}{\partial \eta \partial \zeta}\eta\zeta\right] + \ldots$$

The differentiations are performed at the origin *, *i.e.*, where ξ, η, and ζ all equal zero and where $\Phi_0 = e/r$, with $r = (x^2 + y^2 + z^2)^{1/2}$. The derivatives evaluated on the unit sphere (where $r = 1$), and for a unit charge, represent again the spherical harmonics of Maxwell. The distance coefficients, which include e (contained in Φ) represent the strengths of the corresponding multipoles; for example, ξe represents the x projection of the dipole moment, $\xi^2 e$ is the quadrupole moment in the x direction, etc.

In summary, the potential of a single charge displaced from the origin is equivalent to the sum of potentials of multipoles (including a single pole), all located at the origin. Finally, if we wished to know, at a point x, y, and z, the potential stemming from a number of charges located in some arbitrary manner around the origin of coordinates, we could obtain it by addition of the Taylor series for each charge.

REFERENCES

1. Mossotti, O. F., *Mem. di Mathem. e di Fisica in Modena*, 24, II, (1850) 49.
Clausius, R., *Die Mechanische Wärmetheorie*, Vieweg, 1879, Vol. II, p. 62.
Lorentz, H. A., *The Theory of Electrons*, Dover, New York, 1952; *Ann. Physik*, 9, (1880) 641; *Commun. Acad. Amsterdam*, 1880.
Lorenz, L., *Ann. Physik*, 11, (1880) 70; *Videnskapsselskapets-Skrifter*, 8, (1869) 205; 10, (1875) 485.
2. Onsager, L., *J. Am. Chem. Soc.*, 58, (1936) 1486.

* The coordinates x, y, and z play the roles of parameters which remain constant while we shift attention from the point where the charge is located to the origin.

3. KIRKWOOD, J. G., *J. Chem. Phys.*, 7, (1939) 911; *Ann. N. Y. Acad. Sci.*, 40 (1940) 315.
4. DRUDE, P., *Z. physik. Chem.*, 23, (1897) 267.
5. EINSTEIN, A., *Ann. Physik*, 17, (1905) 549; 19, (1906) 371.
6. PERRIN, J., *Atoms*, transl. by D. L. Hammick, D. van Nostrand Co., New York, 1923.
7. DEBYE, P., *Polar Molecules*, Dover, New York, 1929, Chapter V.
8. PERRIN, J., *loc. cit.*
9. DE BROGLIE, L., *Ann. Physik*, 3, (1925) 22; *Ondes et Mouvements*, Gauthier-Villars, Paris, 1926.
10. SCHROEDINGER, E., *Ann. Physik*, 79, (1926) 361; 79, (1926) 489; 80, (1926) 437; 81, (1926) 109. This series of four articles is reprinted as a volume, *Abhandlungen zur Wellenmechanik*, J. A. Barth, Leipzig, 1927.
11. DEBYE, Ref. 7, p. 145.
12. DEBYE, *ibid.*, Chapter VII.
13. DEBYE, *ibid.*, Chapter X.
14. DEBYE, *ibid.*, Chapter X.

CHAPTER 4

Interaction between Radiation and Matter

The interaction between electromagnetic radiation and matter provides a valuable source of information about the character of a wide variety of materials. Information about the over-all structural characteristics of condensed media such as crystals and porous solids, can be obtained from measurements of such interaction. Particles can also be studied with this kind of measurement: information about size, shape, and configuration can be obtained for particles ranging in size from atoms and molecules to colloids and polymers.

Aside from absorption of radiation energy, a phenomenon that will not be considered here, the presence of matter causes interference with radiation in two ways: (1) there is a slowing down of propagation velocity in the direction of the incident beam ($v = c/\nu$, where ν is the refractive index of the material and c is the velocity of the electromagnetic wave in vacuum) and (2) there is a scattering of radiation in all directions. It is this scattering phenomenon that yields information about the characteristics of particles in the scattering medium.

The phenomenon of scattering can be explained and handled mathematically in two ways. In the more general method the scattering is considered to be caused by inhomogeneities arising from fluctuations in refractive index; Maxwell's equations are applied and the solution fitted to the particular boundary conditions. In the less general but more pictorial method, it is considered that particles under the influence of radiation behave as vibrating dipoles, emitting radiation in all directions; the intensity of the scattered light is calculated through the use of Maxwell's equations. For the most part we shall discuss the light scattering phenomenon in terms of this theory.

Before discussing the interaction of radiation and matter in more detail, we should be aware of the orders of magnitude of electric field strength, particle motion, etc., involved.

References p. 212

According to the 'vibrating dipole' theory, the radiation, proceeding through matter, causes electronic vibrations in the constituent molecules. We can get an idea of the amplitude of the induced electronic vibration through calculation of the field strength present in the incident radiation. We can estimate this for ordinary laboratory light by taking the intensity of strong sunlight as an upper limit. The electric and magnetic intensities of radiation are equal (in vacuum) and the energy density is $E^2/8\pi + H^2/8\pi$. If the radiation is being propagated in the x direction, the energy is related to amplitude A and frequency ω as:

$$\frac{E^2}{8\pi} + \frac{H^2}{8\pi} = \frac{2A^2}{8\pi} \cos^2 (\omega t - kx) \, ,$$

where k is the ratio ω/c. Average energy density is $A^2/8\pi$, since over one period the average of the square of the cosine equals 1/2. Poynting's vector, $\tilde{S}$, representing the current of energy, is simply the average energy density times the velocity of light:

$$\tilde{S} = \frac{A^2}{8\pi} \cdot \tilde{c} \, .$$

The value of the solar constant is 2 cal cm^{-2} min^{-1}, and substituting this value, in proper units, for $\tilde{S}$, we calculate that A equals 1/30 esu or 10 volts per centimeter. The amplitude of electronic vibration induced by a unit field strength (in esu) is of the order of only 10^{-15} cm or 10^{-7} Å,* and so it is apparent that the amplitude of electronic vibration of a dipole under the influence of laboratory radiation is even smaller. We conclude that the effect of interaction of radiation with the electrons in matter proceeds from exceedingly small disturbances; the magnitude of the effect is appreciable only because the charge of the electron is so large.

In considering the magnitude of energy involved, it is convenient to

* In the development of the Cauchy dispersion formula (Chapter 3) we found that the amplitude of electronic vibration turns out to be of the order of the quantity

$$\frac{(e/m)E}{\omega_0{}^2}$$

where ω_0 is the proper frequency of vibration. If we take for $\omega_0/2\pi$ the value 3×10^{15} sec^{-1} (the value we used in Chapter 3) and for m, the mass of the electron, the value 9×10^{-28} gm, then for unit field strength the amplitude is of the order of 10^{-15} cm.

use the photon as the unit of radiation, and discuss energy per photon. This energy varies with frequency; for purposes of comparison we shall use the energy value for a photon of ultraviolet light of 10^{-5} cm wavelength in vacuum, which is 2000×10^{-14} ergs. (To express this energy alternately in terms of voltage, we can use the relation $eV = h\nu$, where e is electron charge, V is voltage in esu, and $h\nu$ is energy per photon. Thus, the voltage is calculated to be 4×10^{-2} esu or 12 international volts.) This energy is large compared with the thermal energy of a molecule: the value of kT at ordinary temperatures is about 4×10^{-14} ergs per molecule. If one wished to reach the energy of such a photon by heating a simple molecule such as helium, the required temperature would be $150{,}000^0$ K, which appears to be a considerable heat in view of the fact that the surface temperature of the sun is about 6000^0 K. The energy of light in this wavelength range is comparable to chemical energies: the energy of a 'mole' of such photons is about 300 kcal, which is of the same order of magnitude as chemical energies. Ultraviolet radiation, then, has an associated energy which is comparable to valence electron energies. The energies associated with radiation in the X-ray region are about a thousand times greater, sufficient to cause disturbances among the inner electrons of the atom. (Conversely, when the inner electrons of the atom are disturbed by, say, bombardment of the atom by high energy electrons, X-ray radiation is emitted.)

We have noted that the energy of radiation varies with wavelength; the intensity of scattered radiation likewise varies with wavelength. This dependency can be deduced by dimensional analysis. Consider a vibrating dipole. A polarization corresponding to the polarizability α is induced by the incoming light. We expect that since the amplitude of the scattered light should depend on the magnitude of the polarizability, then the intensity should depend on α^2. The distance between the point of observation and the scattering center, r, should appear in the denominator of the expression for intensity as r^2, in accordance with the familiar inverse square law. Thus our expression for the ratio of scattered to incident intensity should be $I/I_0 = [\alpha^2/r^2]\, f(\lambda)$. Since the dimensions of α are cm^3, the dimensions of α^2/r^2 are cm^4. It is required that the right-hand side of the equation be dimensionless; the form $f(\lambda) = 1/\lambda^4$ will satisfy this requirement. Our complete expression is:

$$I = I_0\, \alpha^2/r^2\, \lambda^4\,,$$

References p. 212

which is the formula for Rayleigh scattering. This expression shows that the scattering increases greatly with decreasing wavelength*.

The Hertz Vibrating Dipole Theory as Applied to Light Scattering by Particles Small in Comparison with Wavelength; Determination of Molecular Weight

Let us begin our development of light scattering theory by considering, in terms of the vibrating dipole concept, the scattering by particles small in comparison with wavelength.

If a material were absolutely homogeneous, – that is, having uniform density everywhere – no scattering would occur. A gas scatters light because of the inhomogeneities resulting from the concentration of electrons within the molecules. In a liquid solution the difference between the solvent and solute molecules is an additional source of inhomogeneity, and it is the one in which we are interested if we wish to characterize the solute particles. A wavelength can be chosen such that although intermolecular distances between the solute particles are large (compared with the wavelength), the solvent intermolecular distances are so small that the solvent appears practically homogeneous. Thus the 'excess scattering' of solution over pure solvent is close to the total effect: it is the 'excess scattering' that will be considered in our discussions. The chosen wavelength, although small compared to the distances between solute particles, is to be large in comparison with the size of the particle itself, and so we need not consider local differences in scattering by the same particle: the entire particle will appear as a simple point dipole. In addition, if we assume a dilute solution, it is unnecessary to consider any interference of scattered radiation between two or more particles. And as we shall see, the very small fraction of light scattered by a particle makes it unnecessary to consider secondary scattering. Of course, we must pay a price for these simplifications: such a scattering measurement will yield information about the weight of a particle, but not about its specific shape. However, this method is adequate for making the important determination of molecular weight.

A determination of the molecular weight of dissolved particles can be

* This explains, incidentally, the blue color of the sky, for blue light is scattered to a much greater degree than red light.

made by an older method which depends on the measurement of osmotic pressure; however, the light scattering technique has an important advantage for high molecular weight materials. In order to compare the two methods, let us first review the osmotic pressure method. We shall discuss it from a molecular point of view which, although limited in comparison with the thermodynamic approach, affords a pictorial presentation of the process.

Suppose we have a semi-permeable membrane of thickness $\Delta\tau$ separating a water solution of particles from the pure solvent (see Fig. 33).

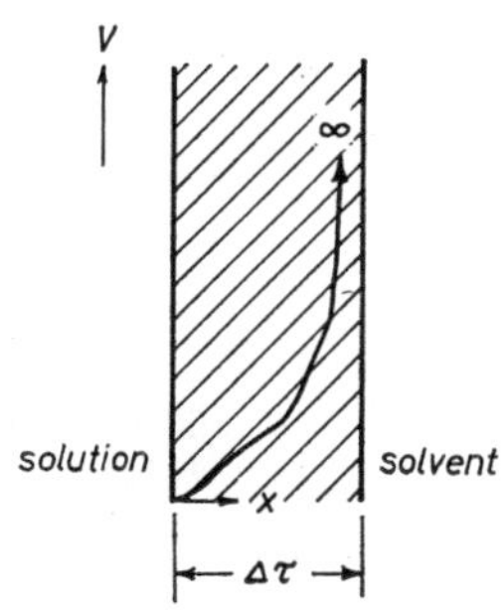

Fig. 33

The membrane is permeable only to the solvent and not to dissolved particles; to say this another way, a particle entering the membrane is subjected to some unknown force, and so has a potential energy which must rise to infinity across the thickness of the membrane since the particle is prevented from passing through. We wish to determine what hydrostatic pressure must be applied to the solution in order to maintain equilibrium – that is, to prevent the pure solvent from either entering or leaving the chamber containing the solution. At equilibrium there is a balance between hydrostatic force and the force associated with the potential energy of the solute particles. Let p = hydrostatic pressure at any point; V = potential energy of a particle at a particular location x within the membrane; f = the force (associated with V) acting on the particle at the location x; and n = number of particles per cubic centimeter. Now let us consider a volume element of unit base area and thickness dx, situated within the membrane in such a way that its base area is parallel to the membrane surfaces. The net hydrostatic force acting on this volume element is $-(\mathrm{d}p/\mathrm{d}x)\,\mathrm{d}x$. The number of particles within the volume is $n\,\mathrm{d}x$, and so the opposing force is $nf\,\mathrm{d}x$. Hence:

References p. 212

$-(\mathrm{d}p/\mathrm{d}x)\,\mathrm{d}x + nf\,\mathrm{d}x = 0$. The force f is related to the potential energy as $-\mathrm{d}V/\mathrm{d}x$, so that we may write: $\mathrm{d}p/\mathrm{d}x + n(\mathrm{d}V/\mathrm{d}x) = 0$. From Boltzmann's law we know that the density of particles is related to the potential energy per particle as $n = n_0\mathrm{e}^{-V/kT}$; therefore our expression may be written: $\mathrm{d}p/\mathrm{d}x + n_0\mathrm{e}^{-V/kT}\,\mathrm{d}V/\mathrm{d}x = 0$. This is the same as $\mathrm{d}p/\mathrm{d}x - kT\,\mathrm{d}n/\mathrm{d}x = 0$, or $\mathrm{d}/\mathrm{d}x\,(p - nkT) = 0$. This is integrated over a path that begins in the solution, goes through the membrane, and ends in the pure solvent. The result is: $p - nkT = p_0$, where p_0 is the hydrostatic pressure in the pure solvent. The excess hydrostatic pressure is the osmotic pressure, π; this is expressed as $\pi = nkT$*, which is Van 't Hoff's law. It usually appears in the more convenient form:

$$\pi = C\frac{RT}{M},$$

where C is the density of solute in grams per cubic centimeter of solution, R is the gas constant Nk (N being Avogadro's number), and M is molecular weight.

The main disadvantage of this method of molecular weight determination is that M appears in the denominator of the expression for π: for a given weight concentration the osmotic pressure decreases with increasing molecular weight. For this reason the method is impractical for really high molecular weight studies, though it is far better than measurements of freezing point depression or boiling point elevation. What is needed for high molecular weight studies is a method which does not decrease in efficiency with increasing molecular weight: methods based on the scattering of radiation meet this requirement.

The scattering method we are considering is for the case of a dilute solution subjected to radiation of wavelength large compared with the particle size of the solute. We consider that the particle solution is equivalent to a solution of dipoles in a uniform solvent medium. The primary light ray will be taken as plane polarized. Also, we assume that the induced moment $\tilde{m}$ lies in the direction of the electric field vector $\tilde{E}$. (This is not true in general, for the particle is not always isotropic.) We shall develop an expression for excess scattering that will include average molecular weight of the particles and so can be used for molecular weight determination by measurement of light scattering.

* The law for the ideal case is obtained because potential energies of interaction between the solute particles were neglected.

In the light scattering phenomenon, we are concerned with the intensity of radiation emitted by a vibrating dipole. Intensity is expressed as energy flux per unit area, which is energy density multiplied by the velocity of propagation of the electromagnetic radiation. In a medium of dielectric constant ε (or ν^2, where ν is refractive index), the density of electric energy is $\nu^2 E^2/8\pi$, as we have already shown. For a sinusoidal electric intensity, such as we have in the case of our vibrating dipole, it is convenient to use the average energy density, which is one-half the maximum value. The magnetic energy density has the same value as the electric energy density, and so the total average energy density is $\nu^2 E_m{}^2/8\pi$, where E_m represents the maximum magnitude of the electric vector.

We can evaluate E_m by means of the expression for the instantaneous electric vector of a vibrating dipole. A development of this expression, which was first derived by Hertz [1], is given in the Appendix. The magnitude of the electric field produced by a vibrating dipole is the magnitude of the real part of Hertz' expression for $\tilde{E}$:

$$E = \mathrm{Re}\left\{\frac{mk^2}{r}\sin\vartheta\, \mathrm{e}^{\mathrm{i}(\omega t - \nu kr)}\right\} = \frac{mk^2}{r}\sin\vartheta\cos(\omega t - \nu kr)\,.$$

Here k is ω/c, a reciprocal length; m is dipole moment; r is distance from the dipole to the point of observation, and ϑ is the polar angle (see Fig. 34). The value of E_m for the scattered radiation is, then:

$$E_m = \frac{mk^2}{r}\sin\vartheta\,.$$

Here m is the maximum value of dipole moment, since it corresponds to the maximum value of the electric field. The quantity $m \sin\vartheta$ is the

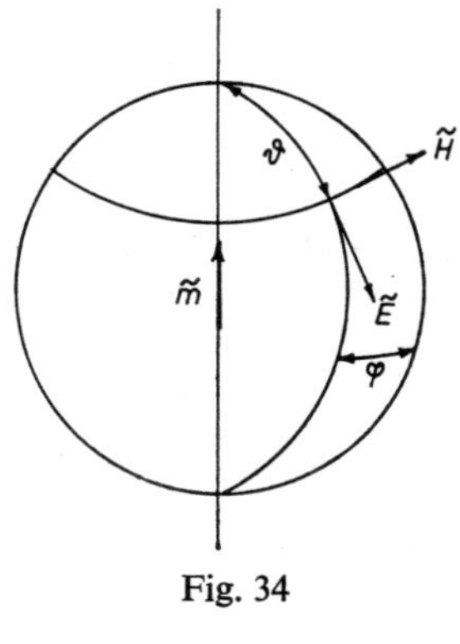

Fig. 34

References p. 212

apparent magnitude of the dipole (in vacuum) as seen by an observer stationed at a point indicated by the polar angle ϑ; the field is a maximum at the equator and falls off to zero at the poles.*

The magnitude of the average energy density of the scattered radiation is accordingly

$$\frac{\nu^2}{8\pi}\frac{m^2k^4}{r^2}\sin^2\vartheta\,.$$

When multiplied by the velocity of light in the medium, c/ν, this gives the magnitude of the flux per unit area of scattered radiation. This quantity is sometimes designated as S because it is the magnitude of Poynting's vector, but more often as I, since it is intensity. Thus:

$$I = \frac{\nu^2}{8\pi}\frac{m^2k^4}{r^2}\sin^2\vartheta\,\frac{c}{\nu}\,.$$

It may be noted that the magnitude of energy density, and consequently of energy flux per unit area varies with r^{-2}. However, if instead of a vibrating dipole we had a static dipole of equal moment, the effect would

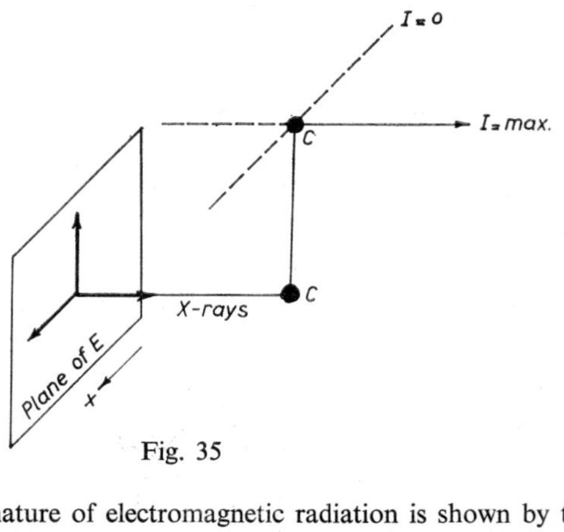

Fig. 35

* The transversal nature of electromagnetic radiation is shown by the results of an experiment performed many years ago with *X*-rays. Referring to Fig. 35 and assuming that the field $\tilde{E}$ is perpendicular to the direction of propagation of the incident ray, we can see that vertical scattering from the first carbon particle would arise from dipoles vibrating with components in the x direction (excited by the non-polarized incident radiation), and would be polarized in that direction. When this beam excites dipoles in the second carbon particle, again no radiation would be sent out toward the poles, and the maximum intensity of radiation would be at the equator. This is the result actually observed.

fall off much more rapidly with distance. This can be seen by inspection of the expression for potential of a static dipole of moment m: since

$$\Phi = \frac{m}{\varepsilon} \frac{\cos \vartheta}{r^2},$$

the magnitude of the electric field strength varies with r^{-3}, and energy density varies with r^{-6}.

For practical purposes, such as the determination of molecular weight, we are interested in the fraction of light scattered. This is the ratio I/I_0, where the subscript 0 refers to incident radiation. For a single particle:

$$\frac{I}{I_0} = \frac{k^4 \sin^2 \vartheta}{r^2} \left[\frac{m}{E_0}\right]^2.$$

(In this equation we need not restrict the value of E_0 to the maximum because the ratio of induced dipole moment to field strength is constant.)

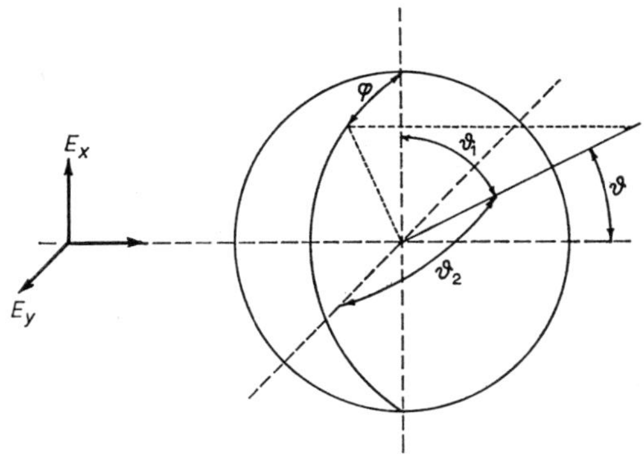

Fig. 36

When the incident light is non-polarized, the energy distribution has a slightly different angular dependence. We may derive the expression for this case simply by modifying the angular dependency. With reference to Fig. 36, we assume that the direction of the incident radiation lies along the z axis with equal electric field components along x and y. Then the total energy flux per unit area in a given direction is proportional to $1/2\ (\sin^2 \vartheta_1 + \sin^2 \vartheta_2)$ which equals $1/2\ (2 - \cos^2 \vartheta_1 - \cos^2 \vartheta_2)$. This can be written in terms of the angle ϑ, the angle measured from the direction of the primary ray. The cosine of the angle between r and x (which is $\cos \vartheta_1$) is $\sin \vartheta \cos \varphi$; similarly, $\cos \vartheta_2$ (the angle between r and y) is $\sin \vartheta \sin \varphi$; and ϑ is the angle between r and z. Therefore, the energy flux per unit area is proportional to $1/2 \left[2 - (\sin^2 \vartheta \cos^2 \varphi + \sin^2 \vartheta \sin^2 \varphi)\right]$ or $(1 + \cos^2 \vartheta)/2$. This term, when used in place of the term $\sin^2 \vartheta$ in the equation for I/I_0, makes the equation applicable when non-polarized light is used. The equation becomes:

$$\frac{I}{I_0} = \frac{k^4}{r^2} \left[\frac{m}{E_0}\right]^2 \frac{1 + \cos^2 \vartheta}{2}.$$

It may be noted that even when the incident radiation is unpolarized, the radiation scattered at right angles to the primary beam is completely polarized, according to our theory. This is because the electric vector is perpendicular to the direction of the incident beam, regardless of the plane of polarization. Of course, it is really the induced vibrating dipole of moment $\tilde{m}$ that causes the scattering of radiation, and if $\tilde{m}$ is not exactly parallel to the electric vector, as is the case in general, then the scattered radiation is not entirely polarized. This is the phenomenon of 'depolarization'.

The above formulas predict that the intensities of radiation scattered by a single particle in the forward and backward directions are the same. Only interferences could produce differences between these intensities, and with our simple model of a dilute solution of 'point molecules' we do not consider interference between scattered rays coming from different molecules or from different parts of the same molecule. Of course, if the disturbing particles were large we would have to consider interferences: then we would expect to find more scattered intensity in the forward than in the backward direction. In fact, measurement of this difference in intensity can be used for a quick estimate of particle size.

Our equation for the scattering of polarized light represents the scattering by a single particle. If now we wish to make the equation applicable for a solution, we must introduce the quantity n (particles per cubic centimeter) and there will be two unknowns in the equation: n and the quantity m/E_0. Therefore we need an independent expression for m/E_0. This quantity, which is the ratio of dipole moment per particle to applied field strength, is the 'excess polarizability' of the particle. We shall designate it as α, though it should be recognized that this excess polarizability, being influenced by the presence of the solvent, is not the same as the polarizability α of a free molecule. The required expression for excess polarizability can be written in terms of dielectric constant, which is measurable through its relation to refractive index. We assume that the particles of solute in a solution of low concentration act as though they were molecules of a gas suspended in a medium of dielectric constant ε_0. It is reasonable, then, to express the change in dielectric constant that results when a small amount of solute is dissolved in the pure medium by use of the law which is valid for a gas:

$$\varepsilon - \varepsilon_0 = 4\pi n\alpha ,$$

where ε is the dielectric constant of the solution. Upon substitution of the square of refractive index for dielectric constant we obtain as a good approximation:

$$\alpha = \frac{m}{E_0} = \frac{\nu^2 - \nu_0{}^2}{4\pi n} = \frac{(\nu - \nu_0)(\nu + \nu_0)}{4\pi n} \cong \frac{2\nu_0(\nu - \nu_0)}{4\pi n} .$$

We now have two equations in our two unknowns. They may be applied to the determination of molecular weight by combining them in an expression for the measurable property of turbidity, which is an aspect of light scattering. The total fraction of light scattered away from the primary polarized beam (of intensity I_0) is:

$$\int_0^\pi \frac{I}{I_0} r^2 2\pi \sin \vartheta \, \mathrm{d}\vartheta ;$$

the integral includes the entire sphere surface. For a single particle this integral equals

$$\frac{8\pi k^4}{3} \left[\frac{m}{E_0}\right]^2 \text{ or } \frac{8\pi k^4}{3} \alpha^2 .$$

Now let us consider a cell of unit cross section and length dx containing n particles per centimeter. The relative change in intensity is:

$$\frac{\mathrm{d}I}{I} = - n \,\mathrm{d}x \cdot \frac{8\pi}{3} k^4\alpha^2 = - \tau \,\mathrm{d}x ,$$

References p. 212

where the quantity τ is excess turbidity and equals

$$\frac{8\pi nk^4}{3}\alpha^2 .$$

(In integrated form the equation is the familiar exponential law: $I = I_0\,e^{-\tau x}$.) The quantity k is, by definition, ω/c or $2\pi/\lambda_v$, where λ_v is the vacuum wavelength of the light. In terms of the wavelength λ in the medium, $k = 2\pi/\nu_0\lambda$. Finally, upon substitution for α and k, excess turbidity is expressed as:

$$\tau = \frac{32\pi^3}{3}\frac{n}{\lambda^4}\left[\frac{(\nu - \nu_0)/\nu_0}{n}\right]^2 .$$

The formula can be written with molecular weight, M, expressed explicitly. This form is obtained by replacing n with CN/M, where N is Avogadro's number and C is weight of solute per cubic centimeter of solution. We may write:

$$\tau = HMC ,$$

where H equals

$$\frac{32\pi^3}{3}\frac{1}{N\lambda^4}\left[\frac{(\nu - \nu_0)/\nu_0}{C}\right]^2 ,$$

and is independent of M.

It should be noted that this development has been for a highly idealized case, that of an infinitely dilute solution of 'point' particles scattering radiation. This means that in order to use this equation in conjunction with experimental measurements, an investigator must find the limit with increasing dilution for the measurements of $(\nu - \nu_0)/C$, which is $\partial\nu/\partial C$, and of turbidity. For $\partial\nu/\partial C$ this is a straight line extrapolation. Turbidity, though, does not yield straight line graphs when plotted against concentration except for the very lowest values. For large molecules, the departure from straight line relationships is pronounced even at low concentrations, a condition that is apparent also in osmotic pressure measurements.

We can now compare the osmotic pressure and turbidity equations involving molecular weight: they are $\pi = RTC/M$ and $\tau = HMC$. At fixed weight concentration, osmotic pressure decreases with increasing molecular weight but turbidity increases. The basic reason for this relationship between turbidity and molecular weight is that for a single molecule the intensity of scattered radiation depends on the square of the induced dipole moment; or, stated equivalently, the relative amount of light scattered depends on the square of the polarizability α. Polarizability is a measure of the volume of a molecule; and since the volume of a polymer molecule, exclusive of trapped solvent, is proportional to the number of monomer units, polarizability should be proportional to the number of monomer units or, equivalently, to the molecular weight of the molecule. For a solution, the intensity of scattered light depends on the magnitude of $n\alpha^2$, which, for a given weight concentration, is proportional to $1/M \cdot M^2$, or M. Thus the efficiency of the method of light scattering increases with molecular weight and so is particularly valuable for study of high molecular weight substances.

There is also a difference between scattering, which is a radiation phenomenon, and refraction, which is an interference phenomenon, in their relationships to solute molecular weight. The change in refractive index upon addition of a small amount of solute to the system depends on the product $n\alpha$, as discussed earlier. The product $n\alpha$ is not dependent on molecular weight, but varies only with weight concentration and the properties of the monomer unit. This means that a given weight percentage of, say, polystyrene in toluene, will produce the same increase in refractive index irrespective of its molecular weight. It is convenient that this is so, for it means that the quantity

$$\frac{(\nu - \nu_0)/\nu_0}{C},$$

which is needed in the light scattering equation, is constant for a given system.

Instead of plotting τ *versus* C, one generally plots HC/τ *versus* C, which should yield $1/M$. A horizontal line would be obtained if all the assumptions of the theory were satisfied, but generally there is some positive slope. At any rate, the intercept at $C = 0$ is taken as M^{-1}. When different solvents are used, the method yields approximately the same values for molecular weight as long as there are no unusual effects due to the solvent, but usually the slopes differ; it is now felt that the magnitude of the slope is a measure of the interaction between solvent and solute molecules.

References p. 212

The order of magnitude of the proportion of light scattered is considerable when molecular weight is high. For example, the addition of polystyrene to benzene gives rise to a $\partial v/\partial C$ value of about 10^{-3}. If the concentration were one percent, the molecular weight were 10^5, and the incident radiation were blue light with a wavelength of 3000 Å, the turbidity would be in the range of 10^{-2}. This is to say that for each centimeter of path travelled, one percent of the incident light would be scattered. This is an easily measured amount.

This method is not as direct as that of osmotic pressure, since in the light scattering procedure one must measure both $\partial v/\partial C$ and turbidity in order to calculate molecular weight. The fact that it makes possible the measurement of very high molecular weight is the strong point of the light scattering method.

Fluctuation Theory and the Determination of Molecular Weight

The scattering of light by gases and dilute solutions can be handled fairly well by the 'vibrating dipole' theory, but this treatment is unsuitable for pure solvents or for concentrated solutions. A pure liquid, for example, exhibits much less scattering than would be expected on the basis of its density alone. Physical interactions between molecules in condensed phases or between dissolved molecules in concentrated solutions, and the resulting large interference effects in the scattered waves, makes the application of the dipole theory entirely impractical. The problem can be solved, but a less graphic idea must be used. By means of the theory of fluctuations, Einstein succeeded in describing the phenomenon quantitatively [2].

For the case of a pure medium, the basic reasoning is as follows. Consider a cube of edge length a, containing the material in question. Let the compressibility of the material be given as κ, the local density as ρ. Suppose at some instant of time there is a particular density distribution within the cube: this can be represented conveniently by a Fourier series:

$$\rho = \rho_0 + \sum_{lmn} A_{lmn} \sin\left[2\pi l\frac{x}{a}\right] \sin\left[2\pi m\frac{y}{a}\right] \sin\left[2\pi n\frac{z}{a}\right],$$

where ρ_0 represents the average density, and magnitudes of the amplitudes A_{lmn} reflect the deviation of the density from the average. It is apparent that such knowledge of density and hence refractive index, as a function of position, would make it possible through Maxwell's equations to calculate the scattering from such a medium. When the calculations are carried through, it turns out that only one quantity, $\overline{A_{lmn}^2}$, is needed to describe the scattering. This result seems reasonable, for one might expect that intensity should depend on the square of an amplitude; and light as seen by the eye is an average intensity.

In order to make a calculation of $\overline{A_{lmn}^2}$, we say that the scattered light observable at any angle is a weighted average taken over all possible configurations of the medium; in other words, the overall scattering picture at any angle is obtained by summing terms, each of which is the product of the intensity of scattered light arising out of a particular density distribution with the probability of the occurrence of that particular distribution. The probability of a particular density distribution may be obtained by use of the statistical method: that is, the ratio of the total energy, U, required to produce a given distribution, to the thermal energy, kT, determines the probability of that state through the familiar expression $e^{-U/kT}$. It seems reasonable (as Einstein shows through a slightly different approach) that one should identify U as the work of compression and dilation necessary to set up the corresponding non-uniform density distribution; this work can be expressed through the A_{lmn}'s and the compressibility κ. With the knowledge of the probabilities corresponding to all possible density distributions, it is possible to calculate the average value, $\overline{A_{lmn}^2}$. This is shown to be (a) independent of lmn and (b) proportional to kT, as one might expect. Since only the quantity $\overline{A_{lmn}^2}$ is required to describe the scattering, the final expression for scattering involves only macroscopically measurable quantities: temperature, a derivative of compressibility, specific volume, and the change in refractive index resulting from a change in density.

Although this theory based on density fluctuations accounts quite well for the scattering phenomena, it must be amplified in order to provide an explanation of the small amount of depolarization that is usually present*. In terms of the vibrating dipole concept, the explanation is

* As has already been discussed, scattered light is completely or almost completely polarized at all angles when the incident radiation is polarized, and at the 90° angle when the incident radiation is unpolarized. The term 'depolarization' refers to the degree to which the scattered light is not polarized under these conditions.

References p. 212

that, in general, molecules are polarized anisotropically; according to fluctuation theory, the explanation is that fluctuations in orientation produce small regular or 'crystalline' regions within the liquid, and these give rise to the depolarization. The proper application of fluctuation theory leads to a good estimate of this effect.

Fluctuation theory may also be advantageously used to calculate the excess amount of scattering produced by a solute. In this case, in place of compression and dilation work, we use work needed to produce fluctuations in concentration. The work required to alter the concentration of solute in a given region is obtained by imagining a semi-permeable membrane surrounding the region, and calculating the osmotic work necessary to produce the concentration change within it. A derivative of osmotic pressure plays the same role for this problem as did a derivative of compressibility in the problem dealing with density fluctuation. Again a Fourier series is employed; here it expresses the fluctuations in concentration. The final expression developed by Einstein involves temperature, specific volume, a derivative of osmotic pressure (instead of a derivative of compressibility), and the familiar quantity $\partial \nu/\partial C$, which is the change in refractive index with concentration (rather than with density). The final expression is:

$$H\frac{C}{\tau} = \frac{\mathrm{d}}{\mathrm{d}C}\left[\frac{P}{RT}\right].$$

Here P is osmotic pressure, and the other functions are defined as in our previous expression for turbidity: τ is excess turbidity; C is concentration; and H is a function of $\mathrm{d}\nu/\mathrm{d}C$, refractive index, and wavelength. For very low concentrations P equals RTC/M, and the formula $HC/\tau = 1/M$ results. This is identical to the equation we developed using the idea of the Hertzian dipole oscillator. The advantage of the more general formula is that, although there is no simple dependence on C, it is applicable to any concentration.

To evaluate the general expression, we approximate osmotic pressure as a Taylor expansion in terms of concentration:

$$\frac{P}{RT} = \frac{C}{M} + BC^2 + \ldots,$$

where B is the second virial coefficient. Then:

$$\frac{\mathrm{d}}{\mathrm{d}C}\left[\frac{P}{RT}\right] = \frac{1}{M} + 2BC + \ldots$$

The coefficient $2B$ represents the slope at the origin of the plot HC/τ *versus* C, and so it can be estimated directly.

It is interesting to contrast the roles played by molecular weight in the two light scattering theories we have discussed. In the vibrating dipole theory, the molecular weight is considered to determine the *size* of the molecule in question, and this in turn determines the magnitude of the induced dipole: there is a clear and obvious connection between molecular weight and amount of scattered light. In the fluctuation theory (as applied to solutions) the connection is not so readily apparent; the dependence of magnitude of fluctuation on molecular weight is less explicit. As shown in Chapter 2, the percentage fluctuation, on the average, in the number of gas molecules to be found in a given cell is inversely proportional to the square root of the average number of particles contained in the cell. The situation for solute molecules is exactly analogous: for a given weight concentration of solute the percentage fluctuation increases as *number* of particles decreases or as molecular weight increases. In summary, in the dipole theory it is the polarizability of the molecule, whereas in the fluctuation theory it is the number concentration of molecules that assumes the major importance.

The Scattering of Radiation whose Wavelength is Not Large Compared with Particle Size

In our discussion of the vibrating dipole theory, we have so far considered the scattering of incident radiation whose wavelength is large compared with particle size. This circumstance enabled us to consider the whole particle as a single point dipole. We were able to relate the intensity of scattered light to the magnitude of the dipole moment, and from this we obtained molecular weight. But if we wish to obtain, in addition, information about the size or even shape of a particle, we must observe it with radiation of a much shorter wavelength. Under such circumstances the particle will not scatter as a whole, for the magnitude of the field strength will vary throughout the particle, and there will be interference

References p. 212

of radiation scattered from various parts of the particle. From an analysis of such a scattering picture it is possible to determine size and perhaps shape of the particle.

This type of study is usually carried out using X-rays. The chief reason for this is that X-rays penetrate most substances in a practically straight line; that is, most substances present to X-rays a refractive index of almost unity. We can demonstrate this fact by making a calculation of refractive index (with respect to X-rays) of a substance; for heuristic purposes we shall do this on the basis of a very simple model. Imagine an electron, bound but free to vibrate, which is subjected to the field of X-rays. We apply the classical equation of motion, which is $m\ddot{x} = -fx - qE\mathrm{e}^{\mathrm{i}\omega t}$, where m is the electron mass, $-fx$ is the restoring force, q is electron charge, and E is the amplitude of the electric field. Because of the very high frequency of X-rays, which is much greater than the proper frequency of vibration, we may neglect the term $-fx$, and describe the motion as $m\ddot{x} = -qE\mathrm{e}^{\mathrm{i}\omega t}$ or $x = (q^2/m\omega^2)\, E\mathrm{e}^{\mathrm{i}\omega t}$. The electric moment, $-qx$, is $-(q^2/m\omega^2)\, E\mathrm{e}^{\mathrm{i}\omega t}$. Hence the electron (so long as the atomic weight of the atom in question is not too high) has for its polarizability in reference to X-rays the value $\alpha = -q^2/m\omega^2$. Polarizability per unit volume, κ, is equal to polarizability per electron multiplied by number of electrons per atom, z, and by number of atoms per unit volume, n; that is, $\kappa = -nzq^2/m\omega^2$. Dielectric constant ε is related to κ as: $\varepsilon = 1 + 4\pi\kappa$, and the refractive index ν equals $\sqrt{\varepsilon}$. Calculations show that this is very close to unity. As an example, we may take carbon, for which the parameters are: $n = 10^{23}\,\mathrm{cm}^{-3}$; $q = 4.8 \times 10^{-10}$ esu; $z = 6$. For X-rays (taking the wavelength to be 10^{-8} cm), the refractive index is calculated to be $1 - (1/2 \times 10^{-6})$. Thus, as a good approximation, we can say that X-rays travel in a straight line through most substances.

It is interesting that the refractive index for X-rays is actually less than unity: the phase velocity in the medium is greater than the velocity of light in vacuum. (Phase velocity is not to be confused with group velocity which carries information.) Accordingly, at certain angles there should be a phenomenon of total reflection as X-rays pass from air to material, and indeed this is observed, but only for grazing angles.

With the use of radiation in the X-ray range for the incident beam, there will occur interferences in the radiation scattered from different points in the scattering particle. Therefore an equation that can be used when there are such interferences must involve phase as well as amplitude of scattered radiation. Let us begin the development of such an equation by considering that we have plane waves incident on a scattering center located at a distance $\tilde{r}$ from an origin of coordinates (see Fig. 37). The parallel lines in the figure represent wave fronts propagated in the direction of the unit vector $\tilde{S}_0$. If A represents the amplitude at the origin at time t, then at the same time the amplitude at the distance $\tilde{r}$ (or at the distance x as measured in the direction of motion of the wavefront) is: $A\mathrm{e}^{\mathrm{i}(\omega t - kx)}$ in the usual complex notation. Again k represents ω/c. This is identical to the expression $A\mathrm{e}^{\mathrm{i}\omega t}\,\mathrm{e}^{-k\tilde{S}_0\cdot\tilde{r}}$, since the scalar product $\tilde{S}_0 \cdot \tilde{r}$ equals $r \cos \vartheta$, or x.

Now, with reference to Fig. 38, suppose the scattering center is located

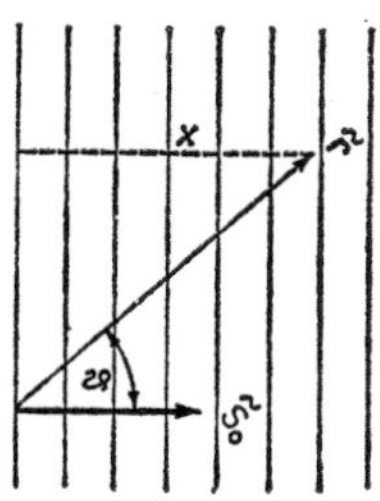

Fig. 37

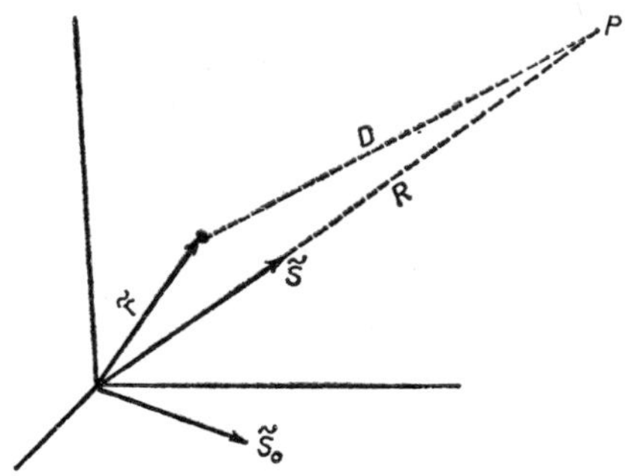

Fig. 38

at the distance r from the origin. What is the amplitude of the scattered wave at a distance R which is very large compared with r? The incident wave is assumed to start from the origin, travel the distance r to the scattering center, and then without change of phase be scattered over the distance D to the observer at point P. The direction of P with respect to the origin is given by the unit vector $\tilde{S}$. If σ represents the scattering factor (that number which characterizes the ability of the particle to scatter radiation), the amplitude at the point P is:

$$A_P = \sigma A\,\mathrm{e}^{\mathrm{i}\omega t}\,\frac{\mathrm{e}^{-\mathrm{i}k\,(\tilde{S}_0\cdot\tilde{r}+D)}}{D}\,.$$

The length D must appear in the denominator because the amplitude of light from a point source varies inversely with distance.

References p. 212

The equation can be simplified by expressing all distances with reference to the origin. Since D and R are approximately parallel, D equals $R - \tilde{S} \cdot \tilde{r}$, and in these terms:

$$A_P = \sigma A \, e^{i\omega t} \, e^{ik(\tilde{S} \cdot \tilde{r} - \tilde{S}_0 \cdot \tilde{r})} \, \frac{e^{-ikR}}{R} \, .$$

Here R replaces D, its near equal, in the denominator where phase is not involved. The factor $e^{ik(\tilde{S} \cdot \tilde{r} - \tilde{S}_0 \cdot \tilde{r})}$ is the only term which depends on the position of the scattering center, and it merely describes the phase of the scattered light. The term e^{-ikR}/R shows that to the observer, the scattered wave (regardless of the position of the center) appears to emanate from the origin. Thus the position of the scattering center determines only the apparent initial phase of the radiation.

When there are a number of scattering centers located around the origin, the observed wave amplitude is the sum of the individual amplitudes. Differences in phase (which arise from differences in position of the scattering centers) produce interferences that result in an observed wave of characteristic amplitude. The equation for the case of n scattering centers may be written (excluding the common factor $e^{i(\omega t - kR)}/R$) as:

$$A_{\text{tot}} = A \sum_n \sigma_n \, e^{ik\tilde{s} \cdot \tilde{r}_n} ,$$

where A_{tot} is proportional to the total amplitude, and $\tilde{s}$ is the vector difference between the unit vectors $\tilde{S}$ and $\tilde{S}_0$. It may be noted that a change in the angle of observation gives rise to a change in $\tilde{s}$ and hence in the amplitude of the scattered wave.

There is a point of great fundamental importance which is assumed in this derivation. It is that the incident ray suffers no appreciable loss in intensity nor change in direction in its passage through the distribution of scattering centers. In other words: (1) it is considered that the initial amplitude acts on each particle, and (2) the vector $\tilde{s}$ is considered to be constant. The very small fraction of light scattered by each particle ensures the validity of point (1) so long as the light path through the sample is not too long. (For exactly the same reason it is unnecessary to consider effects of secondary scattering.) The use of X-rays, which are not appreciably refracted, ensures the validity of point (2).

It has been assumed, in the derivation, that the incident electric vector is polarized in a direction perpendicular to the plane of the paper, and that observations are made in the plane of the paper. This is done so that polar angle, ϑ, remains constant at $\pi/2$.

X-ray Scattering by Crystals

We shall use this equation first to discuss the scattering of X-rays from a structure which is quite large compared with the incident wavelength and which is highly regular: namely, a crystal. The periodic structure of a crystal is characterized by three non-coplanar vectors, $\tilde{a}$, $\tilde{b}$, and $\tilde{c}$, describing a parallelepiped which contains the basic grouping of atoms. Repetition of this parallelepiped, or cell, builds the crystal lattice; the lattice extends indefinitely in all directions. To develop the fundamental picture we may consider the simplest case of one atom per cell, with each atom located at the origin of the three vectors. Let us choose a fixed origin for the entire lattice, and let the position of this origin coincide with the position of an atom. Then in general the position of any atom is given by $\tilde{r}_i = l\tilde{a} + m\tilde{b} + n\tilde{c}$, where l, m, and n are integers. With the knowledge of $\tilde{r}_i$ and the constant vector $\tilde{s}$ we may immediately write the expression for the amplitude of the scattered radiation:

$$A_{\text{tot}} = \sigma A \sum_{lmn} \mathrm{e}^{\mathrm{i}k\,(l\tilde{s}\cdot\tilde{a}+m\tilde{s}\cdot\tilde{b}+n\tilde{s}\cdot\tilde{c})},$$

where the scattering factor σ is assumed constant for all atoms.

The maximum value for amplitude is obtained when the real part of the exponential term is maximal: that is, when $\cos(lk\tilde{s}\cdot\tilde{a} + mk\tilde{s}\cdot\tilde{b} + nk\tilde{s}\cdot\tilde{c})$ is maximal. This occurs when the quantity $(lk\tilde{s}\cdot\tilde{a} + mk\tilde{s}\cdot\tilde{b} + nk\tilde{s}\cdot\tilde{c})$ is an integral multiple of 2π. In order for this to be true, it is necessary that each of the three constituent terms be an integral multiple of 2π. Since l, m, and n are integers, the requirement is: $k\tilde{s}\cdot\tilde{a} = h_1 2\pi$; $k\tilde{s}\cdot\tilde{b} = h_2 2\pi$; and $k\tilde{s}\cdot\tilde{c} = h_3 2\pi$, where the h's are integers. Since $k = 2\pi/\lambda$, an alternate writing for the required condition is:

$$\frac{\tilde{s}}{\lambda}\cdot\tilde{a} = h_1\,;\ \frac{\tilde{s}}{\lambda}\cdot\tilde{b} = h_2\,;\ \frac{\tilde{s}}{\lambda}\cdot\tilde{c} = h_3\,.$$

When this requirement is met, the value of the exponential term in the equation for A_{tot} is unity; then $A_{\text{tot}} = \sigma A$. The interpretation is that

References p. 212

there should be certain values of $\tilde{s}$ for which the observed intensities are maximal. Such maxima are observed, and found to be very sharp.

In solving the three simultaneous equations, we may use the notion of the 'reciprocal lattice'. This is constructed from the physical lattice by means of new vectors, $\tilde{A}$, $\tilde{B}$, and $\tilde{C}$, which are defined as

$$\tilde{A} = \frac{\tilde{b} \times \tilde{c}}{I}, \qquad \tilde{B} = \frac{\tilde{c} \times \tilde{a}}{I}, \qquad \text{and } \tilde{C} = \frac{\tilde{a} \times \tilde{b}}{I},$$

where I is the cell volume $\tilde{a} \times \tilde{b} \cdot \tilde{c}$, or equivalently $\tilde{c} \times \tilde{a} \cdot \tilde{b}$ or $\tilde{b} \times \tilde{c} \cdot \tilde{a}$; hence these new vectors have the dimension $(\text{length})^{-1}$. As may be verified by substitution, the solution may be written:

$$\frac{\tilde{s}}{\lambda} = h_1 \tilde{A} + h_2 \tilde{B} + h_3 \tilde{C}\,.$$

This solution can be used to determine graphically the directions in which radiation maxima occur. The procedure is to construct a reciprocal lattice around an origin based on the three vectors $\tilde{A}$, $\tilde{B}$, and $\tilde{C}$; when the vector $\tilde{s}/\lambda$, laid off from the origin, coincides with a point of the lattice, there will be maximum radiation in that direction. These maxima are sharp because the presence of large destructive interferences acts to reduce the intensity rapidly to zero as one moves away from a direction of maximum intensity.

The scattering of X-rays from a crystal lattice can also be considered from a point of view attributed to Bragg [3]. This viewpoint, which has its own beauty of simplicity, is based on what is sometimes called 'selective reflection'.

A series of parallel equidistant planes is considered to pass through the crystal in such a direction and with such spacing that the atoms are all situated on the planes. (For any crystal there are many possible sets of parallel planes.) When incident light strikes the planes, it will be partially reflected as in an ordinary reflection phenomenon where the angle of reflection equals the angle of incidence. If this were not so, destructive interference would cancel the outgoing light beam.

With reference to Fig. 39, let ϑ be the angle between the directions of incidence and reflection. Because the reflections occur simultaneously from many parallel planes there will be destructive interferences unless the path lengths of adjacent beams differ by a whole number of wavelengths. In other words, in order for there to be a reflection at the angle ϑ, we must have:

$$2d \sin \frac{\vartheta}{2} = n\lambda \,,$$

where n is an integer. This is Bragg's law.

It is easy to see, from Bragg's law, why unique powder patterns are observed for crystalline materials. The expression shows that only for certain angles of incidence will there be light reflected from the crystal planes. For a powdered crystal, where a given set of planes may assume

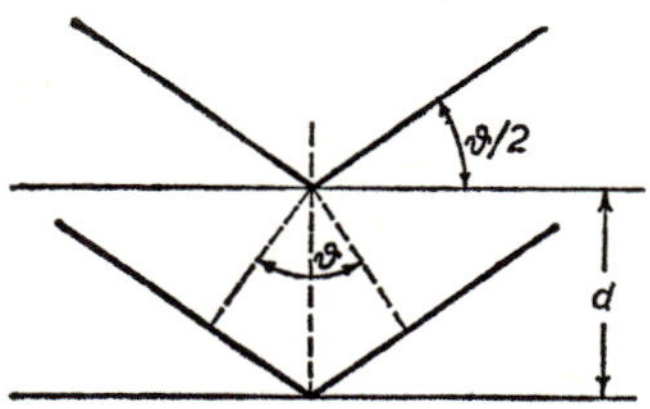

Fig. 39

any configuration around the direction of the primary ray, the rays reflected from these planes lie along the surfaces of cones. A specific crystalline material, with certain values of d, will thus produce cones of characteristic spacing and intensity.

We can combine the results of Bragg's theory and the reciprocal lattice theory into one equation. For this purpose we examine the dependence of the vector $\tilde{s}$ on the angle ϑ. From the definition $\tilde{s} = \tilde{S} - \tilde{S}_0$, we know that $s^2 = S^2 - 2\tilde{S} \cdot \tilde{S}_0 + S_0{}^2 = 2 - 2 \cos \vartheta$, which by means of a trigonometric substitution becomes

$$s^2 = 4 \sin^2 \frac{\vartheta}{2} \,.$$

Hence we may write, upon first squaring the expression for Bragg's law:

$$\frac{n^2}{d^2} = \frac{4 \sin^2(\vartheta/2)}{\lambda^2} = \frac{s^2}{\lambda^2} = h_1{}^2A^2 + h_2{}^2B^2 + h_3{}^2C^2 + 2h_1h_2\tilde{A} \cdot \tilde{B}$$

$$+ 2h_1h_3\tilde{A} \cdot \tilde{C} + 2h_2h_3\tilde{B} \cdot \tilde{C} \,.$$

References p. 212

Up to now we have considered only the 'primitive' lattice structure which contains one atom at each lattice point. Suppose now that there are additional atoms arranged at definite places within each cell. A cell is described as before by the vectors $\tilde{a}$, $\tilde{b}$, and $\tilde{c}$, and within each cell the position of an additional atom ν is given by $p_\nu\tilde{a} + q_\nu\tilde{b} + r_\nu\tilde{c}$; here the parameters p_ν, q_ν, and r_ν are fixed fractions. The position of each atom with respect to the origin is given by $p_\nu\tilde{a} + q_\nu\tilde{b} + r_\nu\tilde{c}$ plus $l\tilde{a} + m\tilde{b} + n\tilde{c}$. The general expression for the amplitude of the scattered light is intrinsically the same as before, but now we must take into account the additional atoms. The expression is:

$$A_{\text{tot}} = A \sum_{lmn,\nu} \sigma_{lmn,\nu}\, \mathrm{e}^{\mathrm{i}k\,(l\tilde{s}\cdot\tilde{a}+m\tilde{s}\cdot\tilde{b}+n\tilde{s}\cdot\tilde{c}+p_\nu\tilde{s}\cdot\tilde{a}+q_\nu\tilde{s}\cdot\tilde{b}+r_\nu\tilde{s}\cdot\tilde{c})}.$$

Because of the invariance of p_ν, q_ν, and r_ν, the directions of the intensity maxima are still governed by the reciprocal lattice construction based on the vectors $\tilde{a}$, $\tilde{b}$, and $\tilde{c}$. Upon substitution of a value of $\tilde{s}$ (obtained from the reciprocal lattice) which gives a direction of maximum intensity, and summing over all the atoms, we obtain:

$$A_{\text{tot}} = A \sum_{lmn,\nu} \sigma_{lmn,\nu}\, \mathrm{e}^{\mathrm{i}\,(lh_1 2\pi+mh_2 2\pi+nh_3 2\pi+p_\nu h_1 2\pi+q_\nu h_2 2\pi+r_\nu h_3 2\pi)}.$$

At the condition of maximum amplitude, the factors $\mathrm{e}^{\mathrm{i}\,lh_1 2\pi}$, $\mathrm{e}^{\mathrm{i}\,mh_2 2\pi}$, and $\mathrm{e}^{\mathrm{i}\,nh_3 2\pi}$ all equal unity, and so:

$$\begin{aligned} A_{\text{tot}} &= A \sum_{lmn,\nu} \sigma_{lmn,\nu}\, \mathrm{e}^{\mathrm{i}\,(p_\nu h_1 2\pi+q_\nu h_2 2\pi+r_\nu h_3 2\pi)} \\ &= A \sum_{lmn} \sigma_{lmn} + A \sum_{\nu} \sigma_\nu\, \mathrm{e}^{\mathrm{i}\,(p_\nu h_1 2\pi+q_\nu h_2 2\pi+r_\nu h_3 2\pi)}. \end{aligned}$$

The last term, which is the summation over the interior atoms, is called the structure factor.

This structure factor shows why some reflections, expected on the basis of the basic lattice structure, do not occur. For example, if the unit cell of the crystal is cubic, with one atom at the origin and one in the center of the cube, then along the direction of a maximum:

$$A_{\text{tot}}/A = \sigma_{lmn} + \sigma_\nu\, \mathrm{e}^{\mathrm{i}\pi\,(h_1+h_2+h_3)}.$$

The first term arises from the atom located at the origin, and the second from the atom in the middle of the cube. If the sum $h_1 + h_2 + h_3$ is an even integer, then $A_{tot}/A = \sigma_{lmn} + \sigma_v$. If the sum is an odd integer, then $A_{tot}/A = \sigma_{lmn} - \sigma_v$, which for identical atoms is zero.

These relationships can be used to obtain values of σ for the two types of atoms in a crystal such as NaCl from observed differences in intensity (which is the square of amplitude). An evaluation of σ can be used, for example, to ascertain the ionic charge of the constituent atoms of a crystal, because σ is roughly proportional to the number of electrons in an atom.

In this consideration of X-ray scattering by crystals, we have assumed that the individual atoms making up the lattice scatter as single independent units. Yet, as we have pointed out, interferences between rays scattered from different parts of the same particle can be neglected only if the incident wavelength is much larger than the particle dimensions, and in general the wavelength of X-rays used in crystal studies is comparable to the size of the atoms. The reason we can neglect such interference is that under certain circumstances such interferences are negligible even when the wavelength and particle length are comparable. Radiation from two points a distance r apart will not interfere significantly in the direction given by our vector $\tilde{s}$ if the term

$$2\pi \frac{\tilde{s}}{\lambda} \cdot \tilde{r}$$

is very small. When $\tilde{s}$ is large (the maximum value is 2, corresponding to backward scattering) and we wish to neglect interferences, our wavelength λ must necessarily be large compared with r. However, if we are able to make our measurements at small angles where $\tilde{s}$ is correspondingly small, then even if λ and r are of comparable size, interference effects will be negligible. The theory we have developed for crystal scattering is, then, valid only for small angles.

For larger angles, the effect of interferences arising from points within the atoms themselves is taken into account by use of an 'atomic form factor' which is characteristic of each kind of atom in the crystal. (At $\vartheta = 0$ the factor reduces to unity.) The factor can be calculated from the known atomic structure and the X-ray diagram predicted, but the reverse procedure, in which one works from the X-ray diagram to the atomic shape, is not easy.

This difficulty has led to a different presentation of the theory. Since the position and density of the electronic cloud rather than the position

References p. 212

of the atomic center is of importance, a Fourier series with unknown coefficients is used to represent the electronic cloud. One then attempts to solve for these coefficients with the help of the information obtained through intensity measurements (generally from a photograph). The problem is complicated by the fact that information concerning the relative phases of scattered radiation is not obtainable from the measurements of intensity.

Light Scattering by Colloids

We have considered the X-ray scattering by crystals from the viewpoint that the scattering structures are large in comparison with the wavelength. The use of low angles, however, has permitted us to assume that the individual atoms making up the lattice scatter as single independent units. Now we shall consider the problem of the scattering at all angles from non-repeating structures with large units – that is, colloidal systems. Because of the absence of order, detailed structural information cannot be obtained; however, the size and shape of the structure can be characterized roughly.

Let us consider first a dilute colloidal suspension of spherical particles of radius a. Only interference effects within individual particles will be considered; the results will be valid for infinite dilution. Experimentally, it is usual to use optical radiation, whose wavelength is comparable to the dimensions of the colloidal particle (as X-ray wavelength is to the size of atoms in crystals). We shall assume that the incident light is plane polarized, the electric vector being perpendicular to the plane of observation so that we are always observing in the equatorial plane*. There is, though, a new difficulty: there may be a large difference between the refractive indices of colloid and solvent, so that it may not be possible to assume that the amplitude and direction of the incident beam are unchanged in passage through a particle in solution. Hence, one aspect of the experimental technique is the attempt to choose a medium with nearly but not quite the same refractive index as the colloid – if there were no difference, there would be no scattering. (In the earliest experiments, proteins were the subject of study, and fortunately the protein

* If unpolarized light is used, a correction factor involving the angle of observation must be included. This was discussed in connection with the Hertzian dipole theory.

structure is so open and filled with solvent that the requirement for small difference in refractive index was met.)

Consider now our colloidal sphere as broken up into small regions, each of volume $d\tau$; every region will contribute to the total scattering a phase factor, $e^{ik\tilde{s}\cdot\tilde{r}}$, and an amplitude factor, σ, which we are going to assume is the same for all regions. The total amplitude, aside from proportionality constants, is given by:

$$A_{\text{tot}} = \sigma \int_V e^{ik\tilde{s}\cdot\tilde{r}}\, d\tau \,.$$

Evaluation proceeds easily with the use of polar coordinates. Let us choose the origin of coordinates to be at the center of the sphere of radius a, and let the fixed $\tilde{s}$ direction be the axis of our polar coordinates, with the polar angle designated as ψ. Then:

$$\frac{A_{\text{tot}}}{\sigma} = \int_0^a \int_0^\pi e^{iksr\cos\psi}\, 2\pi r^2\, dr \sin\psi\, d\psi \,.$$

In integrated form, with α representing the quantity ksa, the equation is:

$$\frac{A_{\text{tot}}}{\sigma} = \frac{4\pi}{3} a^3 \varphi(\alpha) \,,$$

where

$$\varphi(\alpha) = \frac{3}{\alpha^3} (\sin\alpha - \alpha\cos\alpha) \,.$$

This somewhat strange form has been chosen so as to place emphasis on the sphere volume

$$\frac{4\pi}{3} a^3 \,.$$

Because the angle variable $\varphi(\alpha)$ approaches unity when α approaches zero, the amplitude of scattered radiation is proportional to the sphere volume at small angles. Hence intensity of scattered radiation is proportional to the square of the particle volume at small angles. This result can be recognized as a different form of the result we obtained in our expression for the scattering from a 'point' Hertzian dipole: intensity of scattering

References p. 212

from the 'point' dipole was found to be dependent on the square of molecular weight, which is proportional to the square of particle volume. The important point is that for forward scattering, when $s = 0$, the phase factors are all unity, there is no destructive interference of rays emanating from different parts of the same particle, and the molecule may be treated as a 'point' dipole in the classical Hertzian sense. In other words, if data are extrapolated to the zero angle, our Hertzian dipole theory is applicable to particles of any size, including colloids, and will yield information about molecular weight. In order to measure particle size, however, we must rely on destructive interference effects to gain the necessary information. Therefore we need to measure at larger angles where the effect of destructive interference shows up.

The expression for amplitude, as written in terms of σ and $\varphi(\alpha)$, contains all the geometric parameters. A factor containing the square of the difference in index of refraction between particle and solvent is included in the constant σ.

It can be seen by inspection of the expression for $\varphi(\alpha)$, that for non-zero values of α the function goes through zero when $\tan \alpha = \alpha$. (Of course, when $\alpha = 0$, $\varphi(\alpha)$ is unity.) Tan α equals α when α is close to $3\pi/2$, $5\pi/2$, etc., as shown by Fig. 40. The intensity of scattered radiation varies with the square of $\varphi(\alpha)$; Fig. 41 shows roughly the curve of $\varphi^2(\alpha)$

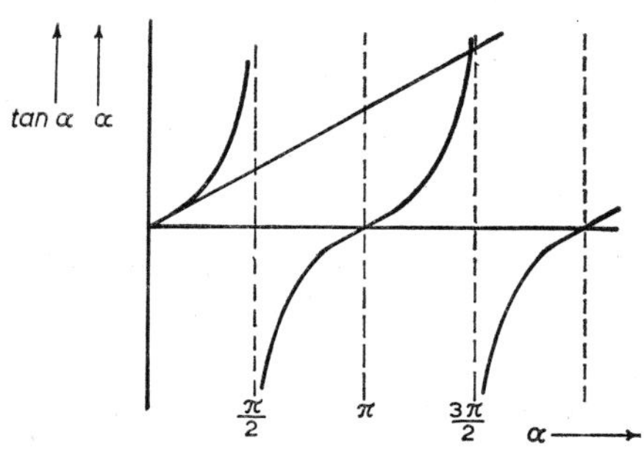

Fig. 40

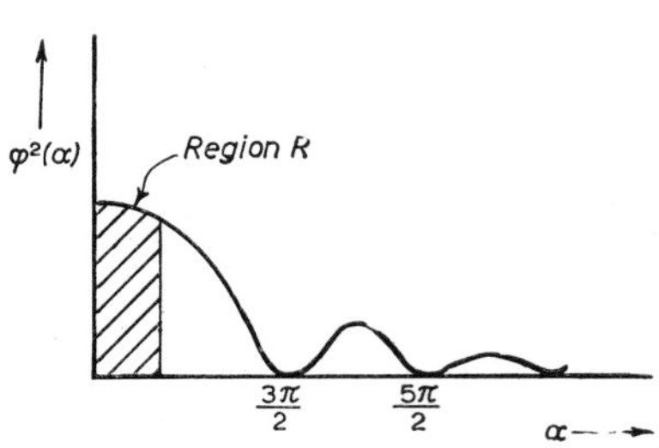

Fig. 41

versus α. Thus the values of ksa at the minima are known, and if the corresponding values of s are determined by observation, then particle radius, a, can be calculated. If the light source is a monochromatic polarized beam, one looks for the directions giving intensity minima. With the use of polarized white light, one looks for colored regions:

for example, in a direction for which the intensity of blue light is a minimum, the beam will be reddish. This provides a good way to determine particle size when the particles are homogeneous.

The question arises: do the parameters of the system allow one to reach a minimum in intensity of scattered light as the angle of observation increases from zero? Because the maximum value of s is 2, the maximum value of α is $2ka$, or $4\pi a/\lambda$. Clearly, if the particle radius a is small compared with the wavelength λ, α will never attain the value $3\pi/2$; observation over the entire angle ϑ would cover only a small range in α, illustrated, for example, by region R in Fig. 41. However, when particles are large in comparison with the wavelength λ, intensity minima are observed.

The term 'minimum of intensity' rather than 'zero intensity' was used, for only in the ideal case of small differences in refractive index where penetration by the light beam is in an almost straight line, is it possible to observe zero intensity. In general, the refractive index differences must be taken into account in a good theoretical calculation. In the particular case of a sulfur sol in water, for example, the refractive index difference is almost 0.2, which is significant. Such calculations have been made by G. Mie[4]. He considered suspensions of spherical particles when the index of refraction differences are large; by fitting the boundary conditions to solutions of Maxwell's equations he obtained the necessary formulas. The calculation is very complicated and will not be dealt with here.

Mie's treatment included even the case of conducting particles: the problem of a sphere with finite conductivity is handled by introducing a complex dielectric constant, the imaginary part involving conductivity. But the problem can be described qualitatively from a more physical point of view. If the electrical field polarizing such a sphere were suddenly removed, the displaced charges would come together and vibrate, emitting strongly damped radiation. Just as in the case of a string under tension, there are natural frequencies of vibration, and the closer the frequency of the incident light is to one of the natural frequencies of vibration, the more the dipole will respond.

There are now available, tables which describe the scattering intensity for spheres of any radius, and for any difference in refractive index between particle and medium. Unfortunately, the problem is so complicated that only for spherical particles is this information available. In practice one tries to have the refractive index difference be as small as possible in order to make use of the simpler theory.

References p. 212

The simpler theory can be used even for non-spherical particles, for by means of the theory, certain intensity measurements can be interpreted so as to yield some general information about the dimensions of the particles. Sufficiently near the observation angle of $\vartheta = 0$, the intensity of scattered radiation is found experimentally to always decrease parabolically like $I/I_0 = 1 - p\vartheta^2$. The identity of the parameter p is established by comparing this empirical formula with a theoretically derived expression for I/I_0. The quantity p turns out to be a measure of the radius of gyration of the particle: specifically, it equals

$$\frac{k^2}{3} r_g^2 ,$$

where r_g is radius of gyration. (This relation, first advanced in a publication by Guinier[5], is discussed more fully in the Appendix.) If the expression for intensity in the neighborhood of the angle $\vartheta = 0$ is further developed as the series $I/I_0 = 1 - p\vartheta^2 + q\vartheta^4 + s\vartheta^6 \ldots$, then q and s are rough measures of the average fourth and sixth powers of r_g. If one assumes that the particles are ellipsoids, then the second and third axes may be obtained, respectively, from q and s – at least in theory. Actually, this has not yet been done.

Light Scattering by Solutions of Finite Concentration

In the limiting case of zero solute (or colloid) concentration, the above development is perfectly adequate. However, as concentration increases, the shape of the curve of intensity *versus* angle of observation changes. The relative intensity (I/I_0) falls off more and more sharply with angle. Eventually, a concentration range may be reached for which the graph increases parabolically near the zero angle, rather than decreases: measurements of relative intensity at small angles show the relation $I/I_0 = 1 + p'\vartheta^2$, instead of $I/I_0 = 1 - p\vartheta^2$. At even higher concentrations, experimental graphs show maxima and minima, as in Fig. 42.

Years ago, when this behavior was first observed, it was thought to reflect a change with concentration of the properties of individual solute particles. But the phenomenon is now attributed to packing; that is, it is considered that at high concentration there is produced around each particle a somewhat regular arrangement of other particles. If we were to focus attention on a single particle and 'move' through the liquid with it, while observing the distribution of particles around it, we could plot the information obtained on a graph which would give the probability of finding the center of another particle at a certain distance *r*. The graph, having maxima and minima, would be similar to that shown in Fig. 43. (There is no probability of finding another particle center within one diameter, 2*a*, of the central particle.) Of course, calculation of such probability would be extremely difficult.

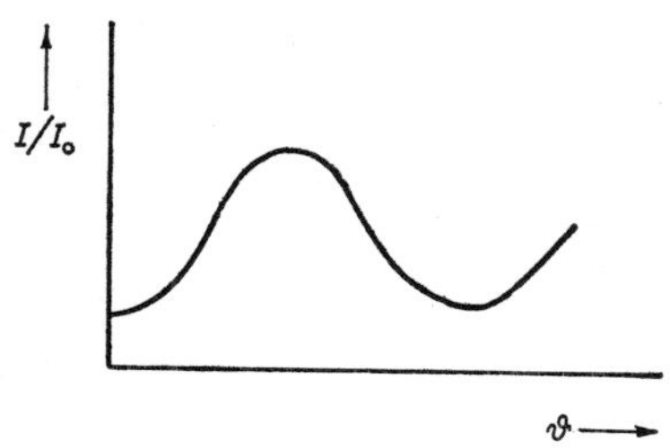

Fig. 42

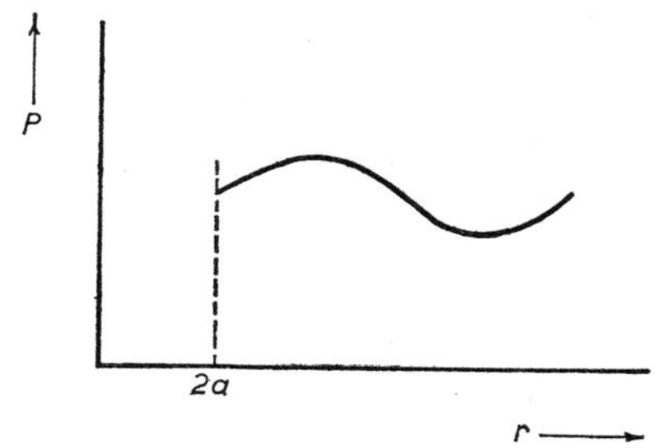

Fig. 43

The first estimation of this probability was obtained by means of a very simple probability experiment designed to reveal order of magnitude only. A covered tray containing a large number of black balls and just two white balls was shaken vigorously, and then the distance between the two white balls measured; a graph presented the outcome of a very large number of trials. This graph turned out to be roughly similar to that in Fig. 43. Later the experimental method was refined to make possible the measurement of separation distance in three dimensional space: spheres were suspended in a liquid, and their spacing, after agitation, recorded photographically. Most recently an evaluation has been made with the use of calculating machines. Analysis was made of a large number of random digits in groups of three, each group representing the location of a particle center; groups which represented overlapping particles were discarded. The probability *versus* distance graph obtained in this way also showed maxima and minima.

With the inclusion of this information, it can be shown that at finite concentration there is a net scattering interference between a given particle and the surrounding ones; thus the whole observed scattering phenomenon is explainable. Since the rigorous calculation is complicated

References p. 212

and not of great significance for our purposes, the handling of intermediate concentrations by the use of a good first approximation will be illustrated.

For this purpose we consider N spherical particles, all of radius a, contained in a volume V. We consider that because of the packing effect, light from one particle will interfere with light from another. As usual, we begin with the vector $\tilde{s} = \tilde{S} - \tilde{S}_0$ that decribes the direction of observation with respect to the primary beam. The incident radiation is taken to be polarized in a direction perpendicular to the plane of the paper and thus for observations taken in the equatorial plane (the plane of the paper), individual dipole scattering is independent of ϑ (or $\tilde{s}$). Some fixed center is chosen within the volume; the distance from this to a given particle center is $\tilde{r}_n$. A volume element $d\tau$ is located by a vector $\tilde{R}_n$ from the fixed center (see Fig. 44). Then in accordance with our

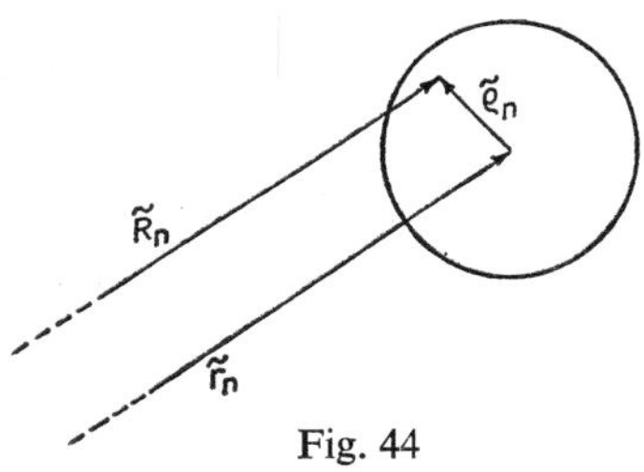

Fig. 44

previously derived equation for amplitude of light scattered by a solution, the amplitude of light scattered by the N particles can be expressed as:

$$A = \sum_{n=1}^{N} \int e^{i k\tilde{s}\cdot\tilde{R}_n} d\tau_n .$$

But $\tilde{R}_n = \tilde{r}_n + \tilde{\rho}_n$; thus:

$$A = \sum_{n=1}^{N} e^{i k\tilde{s}\cdot\tilde{r}_n} \int e^{i k\tilde{s}\cdot\tilde{\rho}_n} d\tau_n .$$

However, as we showed earlier:

$$\int e^{i k\tilde{s}\cdot\tilde{\rho}_n} d\tau_n = \omega\varphi(\alpha) ,$$

where $\varphi(\alpha) = 3/\alpha^3(\sin\alpha - \alpha\cos\alpha)$, and

$$\omega = \frac{4\pi}{3}a^3 .$$

Therefore:

$$A = \omega\varphi(\alpha)\sum_{n=1}^{N} e^{i k\tilde{s}\cdot\tilde{r}_n}.$$

The intensity (for this particular configuration) is given by:

$$I = A^2 = AA^* = \omega^2\varphi^2(\alpha)\sum_{n=1}^{N}\sum_{m=1}^{N} e^{i k\tilde{s}\cdot(\tilde{r}_n-\tilde{r}_m)}.$$

Now let us find the average value for intensity. We designate as $\tilde{r}_{nm}$ the vector $\tilde{r}_n - \tilde{r}_m$, which is the distance between particle centers. It is apparent that when $n=m$ the terms $\tilde{r}_{nm}$ will be zero, each associated exponential will be unity, and there will be N of these. Since there is a total of N^2 terms, this leaves a remainder of $N(N-1)$ terms, each of which has the same average value. Thus we write:

$$< I > = \omega^2\varphi^2(\alpha)\left[N + N(N-1) < e^{i k\tilde{s}\cdot\tilde{r}_{nm}} >\right].$$

So far this has been entirely rigorous. However, for a concentrated solution with much interdependence of position of particles, this average is hard to calculate. As a first approximation (for not too high a particle density) we are going to say that if particle n is fixed, particle m can travel over the entire volume with the exception of that volume occupied by particle n. The average $< e^{i k\tilde{s}\cdot\tilde{r}_{nm}} >$ is then

$$\frac{1}{V}\int_{r=2a}^{R} e^{i k\tilde{s}\cdot\tilde{r}}\,d\tau\ ;$$

$2a$ represents the distance of closest approach between two particles, V is the total volume, and we consider the boundary to be spherical. (This boundary assumption is logical for it can be shown that there is no dependence on macroscopic form of the total volume.) In terms of

References p. 212

known functions, this average is:

$$< e^{i\tilde{k}\tilde{s}\cdot\tilde{r}_{nm}} > = \frac{1}{V}\left\{ \int_0^R e^{i\tilde{k}\tilde{s}\cdot\tilde{r}}\, d\tau - \int_0^{2a} e^{i\tilde{k}\tilde{s}\cdot\tilde{r}}\, d\tau \right\}$$

$$= \varphi(ksR) - \frac{\omega^*}{V}\varphi(2ksa)\ ;\ \omega^* = \frac{4\pi}{3}(2a)^3 .$$

Since for the usual experimental conditions the ratio R/λ is very large, the quantity $\varphi(ksR)$ is negligible at measurable angles†; hence the average is negative. Finally:

$$< I > = \omega^2\varphi^2(\alpha)\left[N - \frac{N(N-1)}{V}\omega^*\varphi(2\alpha) \right],$$

or, in slightly different form:

$$< I > = N\omega^2\varphi^2(\alpha)\,[1 - n\omega^*\varphi(2\alpha)] ,$$

where n is particle density.

This final formula takes into account the effect of interference between rays scattered from different particles as well as between those scattered

† The value of $\varphi(ksR)$, which is

$$\frac{1}{(ksR)^3}[\sin(ksR) - (ksR)\cos(ksR)] ,$$

approaches zero with increasing values of ksR. We have assumed that ksR is large; but at extremely small angles of observation, when s is very small, ksR is not large and $\varphi(ksR)$ is not negligible. How close to the zero angle one would need to measure in order to have $\varphi(ksR)$ be significant can be estimated by a simple calculation for order of magnitude. For instance, for what magnitude of ϑ will the quantity ksR be of the order of unity? If $ksR = 1$, then since $k = 2\pi/\lambda$, and

$$s = 2\sin\frac{\vartheta}{2} \cong \vartheta$$

at low angles, we calculate that

$$\vartheta \cong \frac{1}{2\pi}\frac{\lambda}{R} .$$

Wavelength is of the order of several thousand Ångstroms – say, 3×10^{-5} cm. R is limited by the size of the total volume – if its value were, say, 1 cm, then ϑ would be of the order of only 10^{-5} radians. For all practical purposes, then, it is reasonable to make the assumption that $\varphi(ksR)$ is negligible.

from the same particle. The first term of the equation is N times the intensity expression derived for one particle: this would apply to a situation in which there were no interference between rays scattered from different particles. The second term, then, represents the interference effect arising from finite particle concentration.

The magnitude of this correction term depends, for a given wavelength, on concentration and on the volume ω^*, and under certain circumstances it may be very large. For instance, when there are electric charges which tend to keep the particles apart, the average distance of closest approach may be greater than $2a$, and the volume ω^* may be quite large.

The correction term also has a dependence on the angle of observation. This dependence is implicit in the factor $\varphi(2\alpha)$, since the symbol α represents ksa. For low but finite angles of observation, the value of $\varphi(2\alpha)$ decreases with increasing angle ϑ (this was discussed previously). This is the theoretical explanation of why, at appreciable concentrations, the observed intensity increases rather than decreases with ϑ in the low-angle region.

Characterization of Porous Solids by Light Scattering

The scattering phenomenon has been discussed with respect to two classes of material – the crystal, which is the most highly ordered matter, and the colloidal solution, which has very little order. Detailed information about the structure of a molecule can be obtained through analysis of the pattern of light scattered from its crystal lattice; only general information such as molecular weight and size can be obtained for the colloid in solution. At first it might seem that no useful information at all could be gleaned from light scattering measurements of systems (such as porous solids) in which there is random distribution of electron density, but even in such cases it is possible to characterize the system in some ways.

Suppose that the electron density ρ is some nonperiodic function of distance in the material we want to study. Then with the usual meaning of symbols, the amplitude A at large distances is expressed as:

$$A = \int_V \rho e^{i k\tilde{s}\cdot\tilde{r}} \, d\tau \, .$$

References p. 212

(Again, a proportionality rather than an equality is really correct.) In terms of the average density ρ_{av}, the density at any point is $\rho = \rho_{av} + \Delta$, the quantity Δ being a measure of the density fluctuation. Then, in these terms:

$$A = \rho_{av} \int_V e^{i k\tilde{s}\cdot\tilde{r}} \, d\tau + \int_V \Delta e^{i k\tilde{s}\cdot\tilde{r}} \, d\tau \, .$$

If these two integrals are identified as I and II, then the intensity can be expressed:

$$I = A \cdot A^* = \mathrm{I} \cdot \mathrm{I}^* + \mathrm{I} \cdot \mathrm{II}^* + \mathrm{I}^* \cdot \mathrm{II} + \mathrm{II} \cdot \mathrm{II}^* \, ,$$

where an asterisk refers to the complex conjugate. The term $\mathrm{I} \cdot \mathrm{I}^*$ represents scattering from a material of uniform electron density and for our purposes is unimportant – if the illuminated volume were, say, a cube one millimeter on a side and λ were 10^{-8} cm, then a calculation shows that the angular extent of scattered radiation is $10^{-8}/10^{-1} = 10^{-7}$ radians, which is too close to the primary beam to be observed. Because the average value of the fluctuation is zero, the second and third terms, $\mathrm{I} \cdot \mathrm{II}^*$ and $\mathrm{I}^* \cdot \mathrm{II}$, are also zero †. Thus the intensity is given by the remaining term:

$$I = \mathrm{II} \cdot \mathrm{II}^* = \int \int \Delta_n \Delta_m \, e^{i k\tilde{s}\cdot(\tilde{r}_n - \tilde{r}_m)} \, d\tau_n \, d\tau_m \, .$$

It is the vector $\tilde{r}_n - \tilde{r}_m \equiv \tilde{r}_{nm}$, which connects the two volume elements $d\tau_n$ and $d\tau_m$, that is important (see Fig. 45). If we group together all the

† For example, $\mathrm{I} \cdot \mathrm{II}^*$ equals $\int \int \Delta_n \, e^{ik\tilde{s} \cdot (\tilde{r}_n - \tilde{r}_m)} \, d\tau_n \, d\tau_m$. According to the meaning of the double integral, all possible combinations of $\tilde{r}_n$ and $\tilde{r}_m$ occur in the evaluation of $\mathrm{I} \cdot \mathrm{II}^*$. For our purpose, we shall combine them in a special manner. Suppose for each vector $\tilde{r}_n$, a vector $\tilde{r}_m$ is chosen such that the resultant vector $\tilde{r}_n - \tilde{r}_m$ is constant. In addition, suppose that the differential volumes are all equal. Then the term $e^{ik\tilde{s} \cdot (\tilde{r}_n - \tilde{r}_m)} \, d\tau_m$ will be some constant, C_{nm}. Now the integral $\int_v C_{nm} \Delta_n \, d\tau_n$ is to be evaluated for this one particular value of the vector $\tilde{r}_n - \tilde{r}_m$. It should be noted that the volume over which the integration is to proceed must contain both vectors $\tilde{r}_m$ and $\tilde{r}_n$: this excludes any volume elements $d\tau_n$ for which, although $\tilde{r}_n$ were inside the illuminated volume, $\tilde{r}_m$ would be outside. In other words, the integration volume v is smaller than the illuminated volume V. We are going to assume, however, that the extent in space of each fluctuation is quite small, and that even in this smaller sample of volume, v, the average fluctuation is still zero. Hence $\int_v C_{nm} \Delta_n \, d\tau_n = C_{nm} \int_v \Delta_n \, d\tau_n = 0$. This mathematical process is repeated with different constant vectors $\tilde{r}_n - \tilde{r}_m$ until all the possible combinations of the variables $\tilde{r}_n$ and $\tilde{r}_m$ have been included. Since the result of each integration is zero, then the sum of them, which is $\mathrm{I} \cdot \mathrm{II}^*$, is also zero.

terms $\Delta_n \Delta_m \, e^{i k \tilde{s} \cdot \tilde{r}_{nm}} \, d\tau_n \, d\tau_m$ for which the vector $\tilde{r}_{nm}$ is some constant, and integrate first just these, we obtain:

$$e^{i k \tilde{s} \cdot \tilde{r}_{nm}} \int\int_{\tilde{r}_{nm}=\tilde{C}} \Delta_n \Delta_m \, d\tau_n \, d\tau_m \equiv e^{i k \tilde{s} \cdot \tilde{r}_{nm}} < \Delta_n \Delta_m > \int\int_{\tilde{r}_{nm}=\tilde{C}} d\tau_n \, d\tau_m$$

$$\equiv \int\int_{\tilde{r}_{nm}=\tilde{C}} e^{i k \tilde{s} \cdot \tilde{r}_{nm}} < \Delta_n \Delta_m > d\tau_n \, d\tau_m \, .$$

Strictly speaking, the average $< \Delta_n \Delta_m >$ is a function of the vector $\tilde{r}_{nm}$. But because of the assumption of completely random electron density distribution, it is reasonable to expect that $< \Delta_n \Delta_m >$ should depend only on the magnitude of $\tilde{r}_{nm}$ and not on its direction. If we perform such groupings and integrations until all the possible vectors $\tilde{r}_{nm}$ have been included, and add all the results together, we obtain the desired quantity:

$$I = \text{II} \cdot \text{II}^* = \int_{\tau_m}\int_{\tau_n} < \Delta_n \Delta_m > e^{i k \tilde{s} \cdot \tilde{r}_{nm}} \, d\tau_n \, d\tau_m \, .$$

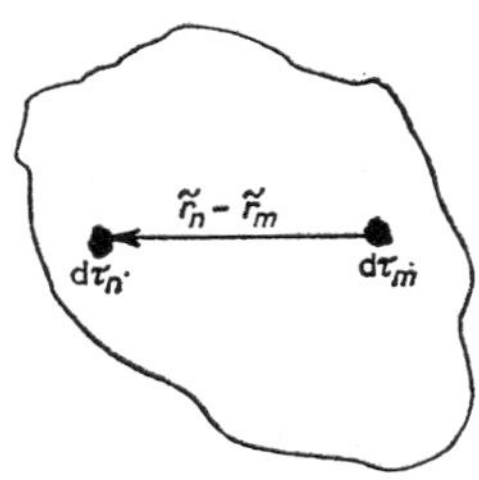

Fig. 45

Appropriate interpretation of this expression provides the desired means of characterizing our material. To make the interpretation, let us suppose that a stick of length r with ends labeled A and B is thrown at random into the material. Then the fluctuation at end A is Δ_A, and at end B is Δ_B. If the product $\Delta_A \Delta_B$ were recorded and the experiment repeated a great number of times, then we would obtain an average, $< \Delta_A \Delta_B >$ which can be recognized as the quantity $< \Delta_n \Delta_m >$. Furthermore, in accordance with our assumption of randomness, this quantity will depend only on the distance r (the absolute magnitude of the vector $\tilde{r}_{AB}$); that is, $< \Delta_B \Delta_A > = f(r)$. For zero stick length, this is just the average of the squares of fluctuation over the volume considered.

References p. 212

For increasingly longer stick lengths, what occurs at one end should depend less and less on what occurs at the other, and $f(r)$ should tend to zero. It is convenient to define a function which we call the 'correlation function' as:

$$\mathrm{C}(r) \equiv \frac{<\Delta_{\mathrm{A}}\,\Delta_{\mathrm{B}}>}{<\Delta^2>}\,.$$

The range of $\mathrm{C}(r)$ is from unity to zero as r varies from zero to infinity; how fast it tends to zero is characteristic of the medium, and this information is what we wish to obtain from an examination of the experimental data.

Using our new terms, we may write:

$$I = <\Delta^2> \iint \mathrm{C}(r)\,\mathrm{e}^{\mathrm{i}\,k\tilde{s}\cdot\tilde{r}_{nm}}\,\mathrm{d}\tau_n\,\mathrm{d}\tau_m\,.$$

Because the correlation function will be appreciably different from zero only for small r, we may assume that there will be a significant contribution to the integral from only a small volume around each point. (Therefore, surface effects may be neglected.) In other words, we may choose some fixed volume element $\mathrm{d}\tau_m$, and consider the expression:

$$\left[\int_{\tau_n} \mathrm{C}(r)\,\mathrm{e}^{\mathrm{i}\,k\tilde{s}\cdot\tilde{r}_{nm}}\,\mathrm{d}\tau_n\right]\mathrm{d}\tau_m\,.$$

This integral should have the same value whatever fixed point $\mathrm{d}\tau_m$ is chosen, and so we may write:

$$\begin{aligned} I &= <\Delta^2> \iint \mathrm{C}(r)\,\mathrm{e}^{\mathrm{i}\,k\tilde{s}\cdot\tilde{r}_{nm}}\,\mathrm{d}\tau_n\,\mathrm{d}\tau_m \\ &= <\Delta^2> \int_{\tau_n} \mathrm{C}(r)\,\mathrm{e}^{\mathrm{i}\,k\tilde{s}\cdot\tilde{r}_{nm}}\,\mathrm{d}\tau_n \int_{\tau_m} \mathrm{d}\tau_m \\ &= <\Delta^2>\,\mathrm{V} \int_{\tau_n} \mathrm{C}(r)\,\mathrm{e}^{\mathrm{i}\,k\tilde{s}\cdot\tilde{r}}\,\mathrm{d}\tau_n\,. \end{aligned}$$

The subscript *nm* has been dropped for convenience. Effectively, we have chosen some origin within the material; $\tilde{r}$ represents the vector drawn from that origin to the volume element $\mathrm{d}\tau_n$, and $\mathrm{C}(r)$ the value of the correlation function for that particular r. We take the polar axis to lie along $\tilde{s}$; expressed in polar coordinates, the volume element $\mathrm{d}\tau_n$ is $2\pi r^2 \sin\vartheta\,\mathrm{d}\vartheta\,\mathrm{d}r$. Thus the quantity $\mathrm{e}^{\mathrm{i}\,k\tilde{s}\cdot\tilde{r}}$ is $\mathrm{e}^{\mathrm{i}\,ksr\cos\vartheta}$. Integrating over the polar angle ϑ, we obtain the final form:

$$I = V < \Delta^2 > \int \mathrm{C}(r) \frac{\sin ksr}{ksr} 4\pi r^2 \,\mathrm{d}r \,{}^*.$$

The correlation function is obtained by a reversal of this Fourier integral. This operation is illustrated by the example of the Fourier integral

$$\varphi(\nu) = 2 \int_0^\infty \mathrm{F}(x) \sin (2\pi\nu x) \,\mathrm{d}x \,,$$

which under the proper conditions of continuity, etc., may be reversed to give

$$\mathrm{F}(x) = 2 \int_0^\infty \varphi(\nu) \sin (2\pi\nu x) \,\mathrm{d}\nu \,.$$

In our particular case let ν be s/λ and x be r; and write ksr, which is $2\pi rs/\lambda$, as $2\pi\nu x$. Then the process of reversal of this integral gives for the correlation function:

$$r\mathrm{C}(r) = \frac{2}{V < \Delta^2 >} \int_0^\infty \frac{s}{\lambda} I\left(\frac{s}{\lambda}\right) \sin\left(2\pi \frac{s}{\lambda} r\right) \mathrm{d}\left(\frac{s}{\lambda}\right).$$

This equation, incidentally, indicates the importance of the ratio s/λ.

Although solved in principle, the problem is difficult to evaluate. One needs the observed value of intensity for all values of s, and this is generally not available. For our purposes the most feasible procedure is to make a guess as to the form of the correlation function, and try this in the expression for intensity.

We know that the correlation function begins at the value unity when the stick length is zero, and decreases to zero with increasing stick length. Furthermore, because we are assuming a random distribution of electron density, it seems reasonable to expect that this decrease should take place in a monotonic fashion, that is, without maxima or minima. Hence, we guess that there is a simple exponential relationship, $\mathrm{C}(r) =$

* It may be noted that intensity as defined here has unusual units, because actually this expression was derived as a proportionality rather than as an equality. In order to express intensity in the usual units, it would be necessary to evaluate the omitted constant. A discussion of this evaluation may be found in a paper by Debye, Anderson, and Brumberger [6].

References p. 212

$e^{-r/a}$, where a is some constant. If this were true, then intensity as a function of s would be:

$$I(s) = \mathrm{V} < \Delta^2 > \int_0^\infty \mathrm{e}^{-r/a} \frac{\sin ksr}{ksr} 4\pi r^2 \, \mathrm{d}r \, .$$

After integration, the equation is:

$$I(s) = \frac{V < \Delta^2 > 8\pi a^3}{\left[1 + (ksa)^2\right]^2} \, * .$$

One may wonder if there is any physical basis for the assumption of an exponential correlation function. A calculation, given in detail in the Appendix, shows that for completely random distributions of void within material such a relationship is obtained.

Although one should not expect that the mathematically simple exponential function will represent the correlation function for every type of random distribution of electron density, there is experimental evidence that it is correct in at least some instances. The intensity equation in which the correlation function appears represents the situation very well for the type of silica gel catalysts used by the petroleum industry in the making of gasoline. A plot of $I^{-1/2}$ *versus* s^2 does yield a straight line as predicted; the slope and intercept of this line should determine the product ka. For values of ksa much greater than unity, the intensity of scattered radiation falls off with $1/s^4$, as predicted.

* The evaluation of the integral proceeds as follows. Let $r/a \equiv \varrho$ and $ksa \equiv \alpha$: hence $r = \varrho a$ and $ksr = \alpha\varrho$. Then the equation is:

$$I = V \langle \Delta^2 \rangle \frac{4\pi a^3}{\alpha} \int_0^\infty \mathrm{e}^{-\varrho} \sin(\alpha\varrho) \varrho \, \mathrm{d}\varrho = - V \langle \Delta^2 \rangle \frac{4\pi a^3}{\alpha} \frac{\mathrm{d}}{\mathrm{d}\alpha} \int_0^\infty \mathrm{e}^{-\varrho} \cos(\alpha\varrho) \, \mathrm{d}\varrho \, .$$

Integration proceeds smoothly in complex notation where $\cos \alpha\varrho = \mathrm{Re}\left\{\mathrm{e}^{\mathrm{i}\alpha\varrho}\right\}$: the complex integral can be evaluated first and then the real part taken. (This follows from the theorem that for a group of complex numbers the sum of real parts is the same as the real part of the sum.) Hence:

$$\mathrm{Re}\left\{\int_0^\infty \mathrm{e}^{-\varrho} \, \mathrm{e}^{-\mathrm{i}\alpha\varrho} \, \mathrm{d}\varrho\right\} = \mathrm{Re}\left\{\frac{1}{1 + \mathrm{i}\alpha}\right\} = \frac{1}{1 + \alpha^2} \, .$$

Then:

$$I = - V \langle \Delta^2 \rangle \frac{4\pi a^3}{\alpha} \frac{\mathrm{d}}{\mathrm{d}\alpha} \left[\frac{1}{1 + \alpha^2}\right] = \frac{V \langle \Delta^2 \rangle \, 8\pi a^3}{[1 + \alpha^2]^2} = \frac{V \langle \Delta^2 \rangle \, 8\pi a^3}{[1 + (ksa)^2]^2} \, .$$

The usefulness of the intensity equation lies in the fact that it makes possible the calculation of the quantity *a* from measurements of intensity as a function of angle of observation. The 'characteristic length' *a* can then be used to calculate surface to volume ratio of a porous material, as will be shown.

The constant *a* has the dimension of length and represents what could be called a 'persistence of electron density'. Figs. 46 and 47 illustrate

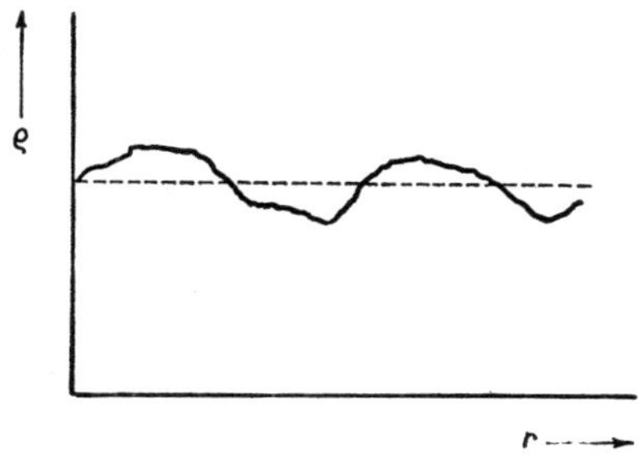

Fig. 46

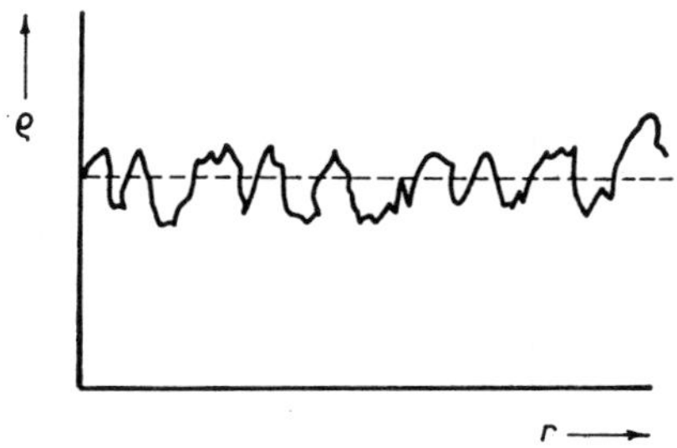

Fig. 47

two different electron density distributions with the same value of $<\Delta^2>$. For case (a) in Fig. 46 the 'persistence of electron density' is greater than for case (b) in Fig. 47, and the important effect on correlation function is illustrated in Fig. 48.

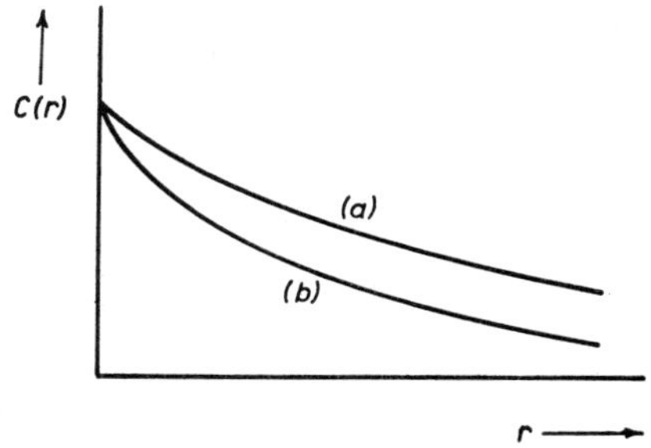

Fig. 48

The value of the parameter *a* in the expression $e^{-r/a}$ is larger in case (a); and so, as can be seen from the equation, the intensity falls off more rapidly with angle ϑ. One has the intuitive feeling that in the case of a catalyst, a small value for *a* should correspond to a large surface area between the gel proper and the pores within the material. This is, in fact, the basis of a light scattering technique for measuring surface area of catalysts which will be discussed shortly.

Before the advent of this light scattering technique, there was already in use a method based on the ability of such a surface to adsorb an inert gas. In this method, the amount of gas (usually nitrogen) adsorbed by a given sample of gel is measured as a function of gas pressure above the gel. These data yield a graph similar to that shown in Fig. 49. The

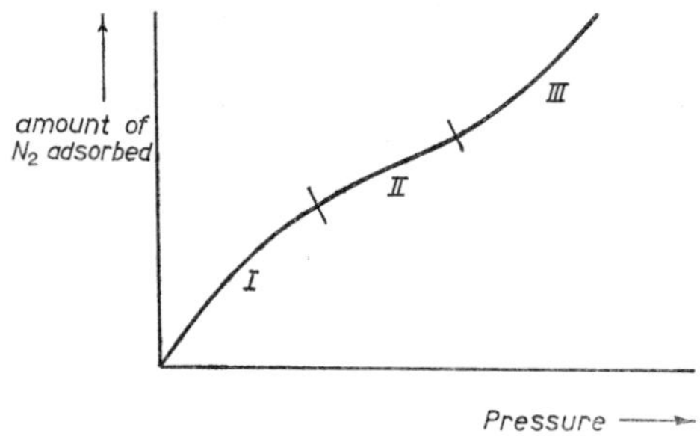

Fig. 49

general interpretation is that region I represents the covering of the gel surface by a monolayer. Succeeding layers are adsorbed with greater difficulty – this is the explanation for region II. Region III represents the liquifaction of nitrogen within the small pores of the gel. With the help of the so-called BET equation [7], which puts this interpretation on a mathematical basis, it is possible to pick out the point on the graph that corresponds to completion of the monomolecular layer; and assuming that the cross section of a nitrogen molecule is known, the surface area of the gel can then be calculated. Unfortunately, there are some poorly understood complications: for instance, when the results of such measurements are compared for two different gels, it is found that the ratio of surface areas is different when carbon tetrachloride is used as the gas than when nitrogen is used. Explanations involving a change in the cross section of the gas molecule have been suggested, but the phenomenon is just not sufficiently understood.

In the light scattering method for measuring surface of a porous gel, we are interested in how the presence of the holes within the material affects the scattering ability of the gel. We shall find it possible (at least in theory) to choose an incident wavelength which is of the order of magnitude of these large inhomogeneities, yet which is large in comparison to atomic spacing, so that for all practical purposes the material proper will appear homogeneous.

A proper wavelength is chosen by considering the relation between intensity of scattered radiation and incident wavelength, as shown in our equation for $I(s)$ in integrated form. Wavelength, λ, appears implicitly in the term ksa (because $k = 2\pi/\lambda$), and from the form of the equation it is apparent that intensity will show significant variation with angle of observation only when ksa is of the order of magnitude of unity. We know that s varies from 0 to 2; therefore we know that λ and a should be comparable in magnitude. An evaluation of a can be made, as will be shown later, from knowledge of surface to volume ratio of the gel: a turns out to be close to the volume to surface ratio. For silica gel, for example, S/V is about 100 m²/cm³, so that V/S is about 100 Å. On this basis we calculate the optimum value for λ to be around 100 Å (which is about 50 times larger than atomic dimensions). Unfortunately, radiation of this wavelength experiences strong absorption and so it is necessary to use X-radiation of a much shorter wavelength, and to observe at correspondingly smaller angles. For this reason the experimental method is usable only at small angles.

In the mathematical treatment of the problem, a great simplification is possible because we need consider only two electron densities, that of the material and that of the hole, rather than a continuously varying electron density. Let $\bar{n}$ be the average electron density of the aggregate of holes and material, and let n be the electron density of the material proper. If φ is the fractional volume of the holes, then $1 - \varphi$ is the fractional volume of material. We have, then:

$$\bar{n} = n(1 - \varphi)\,.$$

In either material or hole, electron density can be expressed as a fluctuation of the average value, $\bar{n}$; that is, as Δ_M or Δ_H. Thus:

$$\Delta_M = n - n(1 - \varphi) = n\varphi\,;$$
$$\Delta_H = 0 - n(1 - \varphi) = -\,n(1 - \varphi)\,.$$

In accordance with our interpretation of the scattering formula, we may define four probabilities relative to density fluctuations encountered at

References p. 212

ends A and B of a stick of length r when it is thrown at random into a sample of the gel. These are called P_{00}, P_{01}, P_{10}, and P_{11}. To illustrate the meaning of these designations, let us consider P_{01} as an example: *if* end A is in a hole, the probability that end B will be in material is P_{01}. In these terms, the quantity $< \Delta_A \Delta_B >$ is readily written as:

$$< \Delta_A \Delta_B > = (1 - \varphi) P_{10} \Delta_M \Delta_H + (1 - \varphi) P_{11} \Delta_M \Delta_M + \varphi P_{00} \Delta_H \Delta_H + \varphi P_{01} \Delta_H \Delta_M .$$

Substituting:

$$< \Delta_A \Delta_B > = -(1 - \varphi) P_{10} n^2 \varphi (1 - \varphi) + (1 - \varphi) P_{11} n^2 \varphi^2 + \varphi P_{00} n^2 (1 - \varphi)^2 - \varphi P_{01} n^2 \varphi (1 - \varphi) .$$

It follows that the average square of fluctuation (corresponding to a stick length of 0) is:

$$< \Delta^2 > = (1 - \varphi) n^2 \varphi^2 + \varphi n^2 (1 - \varphi)^2 = n^2 \varphi (1 - \varphi) .$$

We can now write an expression for the correlation function:

$$C(r) = \frac{< \Delta_A \Delta_B >}{< \Delta^2 >} = -(1 - \varphi) P_{10} + \varphi P_{11} + (1 - \varphi) P_{00} - \varphi P_{01} .$$

Since it is possible to measure φ directly, there are present as unknowns only the four probabilities. From the definitions we see that these probabilities are related: $P_{00} + P_{01} = 1$, and $P_{10} + P_{11} = 1$. And there is a further relationship based on the idea that for a stick of fixed length it is just as likely that end A will be in a hole while end B is in material as it is that the reverse will occur. Since the probability of finding end A within either void or material depends only on the fractional volume, this relationship is $\varphi P_{01} = (1 - \varphi) P_{10}$. Because of the existence of these three relations, it follows that all four probabilities can be expressed in terms of one function of the length of the stick, which we shall call R(r). If we define R(r) such that $P_{01} = (1 - \varphi)$R, then it follows that $P_{10} = \varphi$R, $P_{00} = 1 - (1 - \varphi)$R, and $P_{11} = 1 - \varphi$R. Thus there results a simple relationship between C(r) and R(r):

$$C(r) = 1 - R(r) .$$

This relationship enables us to calculate another probability, the 'probability for dissimilar ends', which leads directly to the calculation of surface to volume ratio. This P_{diss} expresses the chance of finding the two ends in dissimilar surroundings. According to definition:

$$P_{\text{diss}} = \varphi P_{01} + (1 - \varphi) P_{10} ,$$

or, in terms of R:

$$P_{\text{diss}} = 2\varphi(1 - \varphi)\text{R} .$$

Suppose we consider the limiting case for small stick lengths and inquire what the probability is that end A is in a hole and end B is in material. We designate the total volume of the sample as V, and the total interfacial surface as S. The probability that end A will lie within a sheet of thickness dh which is located at a small distance h from the surface is

$$\frac{S\,\mathrm{d}h}{V} .$$

To calculate the probability that end B is simultaneously in material, we consider the sphere, of radius r, whose surface would be described by end B in rotating freely around the position of end A (see Fig. 50).

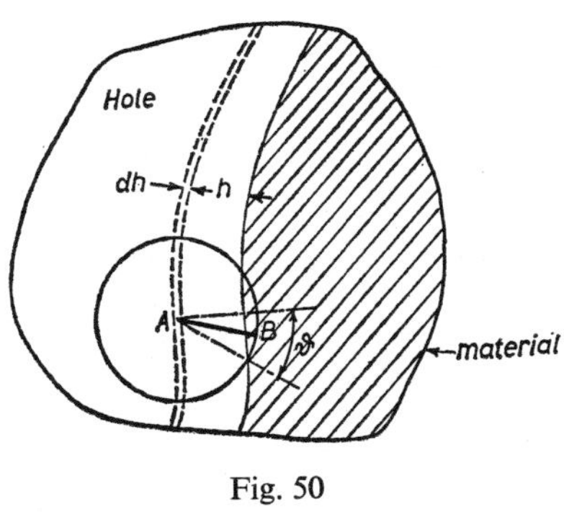

Fig. 50

The probability that end B is in material equals the fractional part of the surface of this sphere that lies within material, for the fixed distance h. (Since we are taking the limiting case of zero stick length, the surface in

References p. 212

the neighborhood of end A may be taken as flat.) This surface fraction is:

$$\frac{1}{4\pi}\int_0^{\vartheta} 2\pi \sin\vartheta \, d\vartheta = \frac{1}{2}(1 - \cos\vartheta) = \frac{1}{2}\left(1 - \frac{h}{r}\right);$$

and so the probability that end A lies within dh and that end B lies within material is the product of the quantities $(1/2)(1 - h/r)$ and $S\,dh/V$. Finally, the probability that end A lies anywhere within the hole while end B lies somewhere in material is the integral:

$$\int_{h=0}^{r} \frac{1}{2}\left(1 - \frac{h}{r}\right)\frac{S}{V}\,dh = \frac{1}{4}\frac{S}{V}r\,.$$

To account for the equally likely situation of A being in material while B is in a hole, we must multiply by 2. Then:

$$P_{\mathrm{diss}} = \frac{1}{2}\frac{S}{V}r\,.$$

Since we are interested in small stick lengths, we may develop $R(r)$ in a Taylor series and keep only the first non-zero term. Since in the limiting case $R(r)$ is zero, the first non-zero term is $R'(0)r$. Thus our first expression for P_{diss} becomes $2\varphi(1 - \varphi)R'(0)r$. Comparison with the other expression for P_{diss} leads to the relation:

$$S/V = 4\varphi(1 - \varphi)R'(0)\,.$$

If it is true that $C(r) = e^{-r/a}$, then since $R(r) = 1 - C(r)$, we can conclude that:

$$S/V = \frac{4\varphi(1 - \varphi)}{a}\,{}^{*}.$$

This is the desired formula relating the surface to volume ratio to the 'characteristic length' a.

---

* For any correlation curve, the tangent at the origin is proportional to the ratio S/V, regardless of the form of $C(r)$.

Before this analysis of surface to volume ratio was available, there existed among investigators, principally in industry, the feeling, well based on experimental results, that the absolute scattered intensity measured at large angles should be a measure of the surface. Using incident radiation of constant intensity, these investigators found that different samples of the same type of catalyst produced at large angles different scattered intensities; those samples which had the largest surfaces as determined by the nitrogen adsorption method, yielded the greatest intensities. This phenomenon may be predicted from our expression for intensity:

$$I(s) = \frac{4\pi V < \Delta^2 >}{\sigma} \int_0^\infty \mathrm{C}(r) \sin(\sigma r) r \,\mathrm{d}r$$

$$= -\frac{4\pi V < \Delta^2 >}{\sigma} \frac{\mathrm{d}}{\mathrm{d}\sigma} \int_0^\infty \mathrm{C}(r) \cos(\sigma r) \,\mathrm{d}r \,,$$

where for convenience we have written σ for ks. The form needed for the consideration of large angle scattering is obtained by a development in powers of $1/\sigma$, most easily accomplished through partial integration. Let $u = \mathrm{C}(r)$ and $\mathrm{d}v = \cos(\sigma r)\,\mathrm{d}r$; then:

$$\int_0^\infty \mathrm{C}(r) \cos(\sigma r) \,\mathrm{d}r = \mathrm{C}(r) \frac{\sin(\sigma r)}{\sigma} \Bigg|_0^\infty - \int_0^\infty \frac{\sin(\sigma r)}{\sigma} \mathrm{C}'(r) \,\mathrm{d}r \,.$$

The first term vanishes. An integration of the second term by the same method yields:

$$\frac{1}{\sigma} \left[\mathrm{C}'(r) \frac{\cos(\sigma r)}{\sigma} \right]_0^\infty - \frac{1}{\sigma} \int_0^\infty \frac{\cos(\sigma r)}{\sigma} \mathrm{C}''(r) \,\mathrm{d}r \,.$$

On substituting limits, we find that the first term reduces to

$$-\frac{1}{\sigma^2} \mathrm{C}'(0) \,;$$

References p. 212

partial integration of the second term and subsequent substitution of limits leads to terms in $1/\sigma^3$ and higher. Since we are interested in the limiting case of large σ we shall retain only the term in $1/\sigma^2$. Using this value in the expression for intensity, we obtain the form:

$$I(s) = \frac{8\pi V < \Delta^2 >}{k^4 s^4} \mathrm{R}'(0).$$

We know that $< \Delta^2 > = n^2 \varphi (1 - \varphi)$ and

$$\mathrm{R}'(0) = \frac{1}{4\varphi(1 - \varphi)} \frac{S}{V};$$

hence the final form, which shows the dependence on surface, may be written:

$$I(s) = \frac{2\pi n^2}{k^4 s^4} S.$$

This formula, which is valid for large angles only, provides a means of measuring the total surface illuminated. The formula derived for the case of small angles of observation gives a measure of surface to volume ratio.

It can be seen that because of the presence of the factor n^2, representing the square of the electron density of the material proper, the ratio of large angle intensities and the ratio of surface areas should be the same only when catalysts having the same electron density in their material are compared. This is what had been deduced empirically by industrial investigators.

Critical Opalescence

A recent use of correlation function has been in the study of the phenomenon of critical opalescence[8]. Under certain experimental conditions, ordinarily small molecules may form large aggregates of size comparable to the wavelength of light. Measurements of light scattering by such systems have been related theoretically to a characteristic distance interpreted as the range of molecular interaction.

APPENDIX TO CHAPTER 4

(*i*) *The Vibrating Electric Dipole*

The calculation of the electromagnetic energy radiated by an oscillating electric dipole was first performed by Hertz. We shall present here the highlights of a development of the expression *.

The calculation is based on the Maxwell equations, which can be written:

(1) $$\text{curl}\,\tilde{E} + \frac{\mu}{c}\frac{\partial \tilde{H}}{\partial t} = 0\,;$$

(2) $$\text{curl}\,\tilde{H} - \frac{\varepsilon}{c}\frac{\partial \tilde{E}}{\partial t} = 4\pi\,\tilde{J};$$

(3) $$\text{div}\,\tilde{H} = 0\,;$$

(4) $$\text{div}\,\tilde{E} = \frac{4\pi}{\varepsilon}\rho\,.$$

The vectors $\tilde{E}$ and $\tilde{H}$ are, respectively, the intensities of the electric and magnetic fields, c is the velocity of an electromagnetic wave in vacuum, μ is magnetic permeability, ε is dielectric constant, $\tilde{J}$ is current density, and ρ is charge density. These quantities have been written in Mixed units.

Maxwell's equations express, in differential form, well known experimental laws. The first expresses Faraday's law which concerns the production of electromotive force by a changing magnetic field. The second equation is based on Ampere's law concerning the magnetic field produced by an electric current. The term

$$\frac{\varepsilon}{c}\frac{\partial \tilde{E}}{\partial t}$$

* For a more complete discussion of electromagnetic theory and its application to this problem, one can refer to a standard text such as Stratton[9].

References p. 212

represents Maxwell's 'displacement current' and was introduced by him in order to make all the equations consistent with the idea of the conservation of charge: div $\tilde{J} + \frac{1}{c}\partial\rho/\partial t = 0$. It was the introduction of this term that made it possible to combine equations (1) and (2) to yield an expression that describes the propagation of a wave – the 'wave equation'. The last two equations express the absence of free magnetic charge, and the dependence of the electric field on the density of charge. For the purpose of our calculation we assume the absence of magnetic material, so that μ is unity, and also the absence of current and charge, so that $J = 0$ and $\rho = 0$.

In dealing with electrostatics, it is easier to introduce the notion of a scalar potential, Φ, than it is to attempt to solve the vector differential equations directly. Unfortunately, since in the general Maxwell equations curl $\tilde{E}$ does not necessarily vanish, grad Φ alone is not a solution for the electric field intensity. Other general vector properties, however, may be used in an analogous manner to simplify these equations. In particular, since the operation div curl yields zero identically, and since div $\tilde{H} = 0$, we can put $\tilde{H} = \text{curl } \tilde{A}$. Substitution into the first Maxwell equation yields:

$$\text{curl } \tilde{E} + \frac{1}{c}\frac{\partial}{\partial t}\text{curl } \tilde{A} = \text{curl}\left(\tilde{E} + \frac{1}{c}\frac{\partial \tilde{A}}{\partial t}\right) = 0\,.$$

The quantity $-$ grad Φ can now be introduced to represent this new vector whose curl vanishes everywhere:

$$-\text{ grad } \Phi = \tilde{E} + \frac{1}{c}\frac{\partial \tilde{A}}{\partial t}\,,$$

or, equivalently:

$$\tilde{E} = -\text{ grad } \Phi - \frac{1}{c}\frac{\partial \tilde{A}}{\partial t}\,.$$

The second term of this equation is absent in electrostatics. The second and fourth Maxwell equations become:

$$\text{curl curl } \tilde{A} + \frac{\varepsilon}{c}\text{ grad }\frac{\partial \Phi}{\partial t} + \frac{\varepsilon}{c^2}\frac{\partial^2 \tilde{A}}{\partial t^2} = 0\,;$$

$$\operatorname{div} \operatorname{grad} \Phi + \frac{1}{c} \operatorname{div} \frac{\partial \tilde{A}}{\partial t} = 0 .$$

As they stand, these two equations each contain Φ and $\tilde{A}$. However, by finding a suitable relation between these quantities, we are able to separate them into two independent equations. This is done in the following way. If to the function $\tilde{A}$, we add another function ($-$ grad ψ) which is arbitrary and whose curl identically vanishes, then the physically significant function, $\tilde{H}$, will be unchanged. This is so because curl $\tilde{A}_0 \equiv \operatorname{curl}(\tilde{A} - \operatorname{grad} \psi) = \operatorname{curl} \tilde{A} = \tilde{H}$. In other words, $\tilde{A}$ is not a unique function. Substitution of the new function $\tilde{A}_0$ into the first Maxwell equation yields a new value, Φ_0, which differs from the old by the quantity

$$\frac{1}{c} \frac{\partial \psi}{\partial t} .$$

That is:

$$\tilde{A} = \tilde{A}_0 + \operatorname{grad} \psi ;$$

$$\Phi = \Phi_0 + \frac{1}{c} \frac{\partial \psi}{\partial t} .$$

If we take the divergence of both sides of the first equation, and to each side add the time derivative of the second equation, multiplied by ε/c, we obtain:

$$\operatorname{div} \tilde{A} + \frac{\varepsilon}{c} \frac{\partial \Phi}{\partial t} = \operatorname{div} \tilde{A}_0 + \frac{\varepsilon}{c} \frac{\partial \Phi_0}{\partial t} + \operatorname{div} \operatorname{grad} \psi + \frac{\varepsilon}{c^2} \frac{\partial^2 \psi}{\partial t^2} .$$

Because the function ψ is arbitrary, one can choose it so as to make the entire right-hand side vanish; then:

$$\operatorname{div} \tilde{A} = - \frac{\varepsilon}{c} \frac{\partial \Phi}{\partial t} .$$

This relation facilitates the mathematical operations without influencing the calculated values of the physically significant quantities $\tilde{E}$ and $\tilde{H}$. If this is substituted into the second and fourth Maxwell equations (written in terms of $\tilde{A}$ and Φ) we obtain:

References p. 212

$$\text{curl curl } \tilde{A} - \text{grad div } \tilde{A} + \frac{\varepsilon}{c^2}\frac{\partial^2 \tilde{A}}{\partial t^2} = 0\ ;$$

$$\text{div grad } \Phi - \frac{\varepsilon}{c^2}\frac{\partial^2 \Phi}{\partial t^2} = 0\ *.$$

This is the separated and most convenient form of the Maxwell equations: each equation describes the propagation of a wave in space.

We are concerned only with periodic radiation – in particular, with sinusoidal radiation. Hence $\tilde{E}$, $\tilde{H}$, $\tilde{A}$, and Φ will be proportional to $e^{i\omega t}$. The time derivatives are, then:

$$\frac{\partial \tilde{A}}{\partial t} = i\omega\tilde{A}\ ; \qquad \frac{\partial \Phi}{\partial t} = i\omega\Phi\ ; \qquad \frac{\partial^2 \tilde{A}}{\partial t^2} = -\ \omega^2\tilde{A}\ ; \qquad \frac{\partial^2 \Phi}{\partial t^2} = -\ \omega^2\Phi\ .$$

With the substitution of these expressions, and the use of the vector identity curl curl $\tilde{A}$ – grad div $\tilde{A}$ = – div grad $\tilde{A}$ † we obtain:

$$\text{div grad } \tilde{A} + \nu^2 k^2 \tilde{A} = 0\ ;$$

$$\text{div grad } \Phi + \nu^2 k^2 \Phi = 0\ ,$$

where k is defined as ω/c and the refractive index $\nu = \sqrt{\varepsilon}$; hence

$$\nu k = \frac{\nu\omega}{c} = \frac{2\pi}{\lambda}\,,$$

where λ is the wavelength of the electromagnetic wave in the medium of refractive index ν.

The problem at hand concerns the radiation from an oscillating dipole. In terms of mathematics, this means that we seek a solution for the 'vector potential' $\tilde{A}$ that goes to infinity like $1/r$ at the location of the dipole – this is the simplest electromagnetic case, just as in electrostatics

* From the form of the wave equation it is observed that the velocity of propagation in the medium is given by $c/\sqrt{\varepsilon}$. Thus the refractive index ν, appears as $\sqrt{\varepsilon}$. This famous relationship is attributed to Maxwell.

† This does not hold if the components of $\tilde{A}$ are expressed in other than Cartesian coordinates; see Stratton [9] for a clear discussion of this point.

the point charge is the simplest source and gives rise to a potential proportional to $1/r$.

We let the dipole be oriented along the z axis. Because of symmetry, we set $A_x = A_y$, and furthermore assume that they each equal zero (the validity of this assumption will become apparent). Being guided by the simplest electrostatic case, we seek a spherically symmetrical solution to A_z; in other words, we look for the solution to the equation:

$$\frac{1}{r^2}\frac{\mathrm{d}}{\mathrm{d}r}\left(r^2\frac{\mathrm{d}A_z}{\mathrm{d}r}\right) + \nu^2k^2A_z = 0\,,$$

where div grad A_z has been expressed in spherical coordinates. (The polar angle ϑ does not appear in this equation but will be required for subsequent expressions in terms of these spherical coordinates.) Such a solution is:

$$A_z = C\frac{\mathrm{e}^{-\mathrm{i}\nu kr}}{r}\,.$$

The negative exponent is chosen because the waves progress from the origin to infinity. Hence the total solution, omitting the time factor $\mathrm{e}^{\mathrm{i}\omega t}$, is:

$$\tilde{A} = C\frac{\mathrm{e}^{-\mathrm{i}\nu kr}}{r}\tilde{k}\,,$$

$\tilde{k}$ being the unit vector in the direction of the z axis.

The constant C may be evaluated by considering the situation in which the wavelength λ is much greater than the distance r. Under these conditions, the potential at the distance r is equivalent to the static case for potential due to a dipole; this is $\Phi = \tilde{m}\cdot\tilde{r}/\varepsilon r^3$, where m refers to dipole moment. For our case, $\tilde{m}\cdot\tilde{r} = mr\cos\vartheta = mz$, so that:

$$\Phi = \frac{mz}{\nu^2r^3}\,.$$

We can write another expression for Φ for the condition

$$kr = 2\pi\frac{r}{\lambda} \ll 1\,.$$

References p. 212

We begin with the relationship between $\tilde{A}$ and Φ, which we have expressed as

$$\operatorname{div} \tilde{A} = -\frac{\varepsilon}{c}\frac{\partial \Phi}{\partial t}.$$

This is equivalent to div $\tilde{A} = -\,\mathrm{i}\nu^2 k\Phi$. Thus $\Phi = -\operatorname{div} \tilde{A}/\mathrm{i}\nu^2 k$. Also, we know that in our case

$$\operatorname{div} \tilde{A} = \frac{\partial A_z}{\partial z};$$

and since

$$\frac{\partial A_z}{\partial z} = \frac{\mathrm{d}A_z}{\mathrm{d}r}\cdot\frac{\partial r}{\partial z} = \frac{\mathrm{d}A_z}{\mathrm{d}r}\cdot\frac{z}{r},$$

then

$$\operatorname{div} \tilde{A} = \frac{\mathrm{d}A_z}{\mathrm{d}r}\cdot\frac{z}{r} = -\,C\,\mathrm{e}^{-\mathrm{i}\nu kr}(1 + \mathrm{i}\nu kr)\frac{z}{r^3}.$$

Thus:

$$\Phi = C\,\mathrm{e}^{-\mathrm{i}\nu kr}\frac{z}{r^3}\cdot\frac{\nu kr - \mathrm{i}}{\nu^2 k}.$$

When $kr \ll 1$, this reduces to:

$$\Phi = \frac{C}{\mathrm{i}\nu^2 k}\frac{z}{r^3}.$$

By comparison of our two expressions for Φ, we can establish the value of the constant C:

$$C = \mathrm{i}km\,.$$

Thus the final expressions for Φ and $\tilde{A}$ (exclusive of the time factor) are:

$$\Phi = \frac{m}{\nu^2}(1 + \mathrm{i}\nu kr)\,\mathrm{e}^{-\mathrm{i}\nu kr}\frac{z}{r^3};$$

$$\tilde{A} = \mathrm{i}km\frac{\mathrm{e}^{-\mathrm{i}\nu kr}}{r}\tilde{k}\,.$$

These expressions are valid for all distances; however, for the practical case of light scattering, our interest lies in observing, at macroscopic distances, light with a wavelength of about 5×10^{-5} cm that has been scattered by a molecule. This means that we are interested in the range

$$kr = 2\pi \frac{r}{\lambda} \gg 1 .$$

In this range our equations become (exclusive of the time factor):

$$\Phi = \frac{ikm}{v} \, \mathrm{e}^{-\mathrm{i}\nu kr} \frac{z}{r^2} ;$$

$$\tilde{A} = ikm \frac{\mathrm{e}^{-\mathrm{i}\nu kr}}{r} \tilde{k} .$$

In order to solve the problem, it is now only necessary to construct the vectors $\tilde{E}$ and $\tilde{H}$, representing the electric and magnetic components of the electromagnetic radiation. As discussed earlier, these field quantities are related to the scalar and vector potentials Φ and $\tilde{A}$:

$$\tilde{E} = -\,\mathrm{grad}\,\Phi - \frac{1}{c}\frac{\partial \tilde{A}}{\partial t} ;$$

$$\tilde{H} = \mathrm{curl}\,\tilde{A} .$$

The most convenient way to express the operators grad and curl, is to use polar coordinates. Thus for grad Φ we write:

$$\mathrm{grad}\,\Phi = \frac{\partial \Phi}{\partial r} \tilde{i}_1 + \frac{1}{r}\frac{\partial \Phi}{\partial \vartheta} \tilde{i}_2 ,$$

where $\tilde{i}_1$ and $\tilde{i}_2$ are the mutually perpendicular unit vectors that point, respectively, in the directions of increasing r and ϑ. (In our problem, there is no dependence on the azimuthal angle.) With proper substitution ($z = r \cos \vartheta$) and manipulation, the operation yields:

$$-\,\mathrm{grad}\,\Phi = \frac{ikm}{v} \frac{\mathrm{e}^{-\mathrm{i}\nu kr}}{r} \left[\left(i\nu k + \frac{1}{r} \right) \cos \vartheta \, \tilde{i}_1 + \frac{\sin \vartheta}{r} \tilde{i}_2 \right] .$$

References p. 212

If the condition $kr \gg 1$ is again employed, the equation may be reduced to:

$$-\operatorname{grad} \Phi = \frac{ikm}{v}\frac{e^{-i\nu kr}}{r}\left[i\nu k \cos \vartheta\, \tilde{i}_1 + \frac{\sin \vartheta}{r}\, \tilde{i}_2\right].$$

This represents the first term in the expression for $\tilde{E}$; the second term is evaluated very simply. Using the evaluation $\partial \tilde{A}/\partial t = i\omega \tilde{A}$, we write:

$$-\frac{1}{c}\frac{\partial \tilde{A}}{\partial t} = -\frac{i\omega}{c}\, ikm \frac{e^{-i\nu kr}}{r}\, \tilde{k}\,.$$

The polar expression for this vector is obtained by use of the fact that the unit vector $\tilde{k}$ can be written as $\tilde{k} = \cos \vartheta \tilde{i}_1 - \sin \vartheta \tilde{i}_2$. In these new coordinates:

$$-\frac{1}{c}\frac{\partial \tilde{A}}{\partial t} = k^2 m \frac{e^{-i\nu kr}}{r} \cos \vartheta \tilde{i}_1 - k^2 m \frac{e^{-i\nu kr}}{r} \sin \vartheta \tilde{i}_2\,.$$

When the two terms in the expression for $\tilde{E}$ are combined, the radial components cancel (demonstrating that the vector $\tilde{E}$ is transverse to the direction of propagation). The result, with the time factor $e^{i\omega t}$ included, is:

$$\tilde{E} = -\, mk^2 \frac{e^{i(\omega t - \nu kr)}}{r} \sin \vartheta\, \tilde{i}_2\,.$$

The evaluation of curl $\tilde{A}$, which gives the magnetic field intensity $\tilde{H}$, proceeds in a similar way. The vector operator curl is first expressed in polar coordinates. Thus:

$$\operatorname{curl} \tilde{A} = \frac{1}{r}\left[\frac{\partial}{\partial r}\left(rA_2\right) - \frac{\partial A_1}{\partial \vartheta}\right] \tilde{i}_3\,,$$

for our case in which there is no dependence on the azimuthal angle, and no component in the azimuthal direction. Here $\tilde{i}_3$ is the unit vector in the azimuthal direction, and A_1 and A_2 are the components along the directions described by $\tilde{i}_1$ and $\tilde{i}_2$. We can find the components A_1 and A_2

by substituting into our equation for $\tilde{A}$ the vector identity of $\tilde{k}$ (which is $\cos\vartheta\,\tilde{\imath}_1 - \sin\vartheta\,\tilde{\imath}_2$) and making the identification directly. The equation for $\tilde{A}$ becomes:

$$\tilde{A} = ikm\left[\frac{\mathrm{e}^{-\mathrm{i}\nu kr}}{r}\cos\vartheta\,\tilde{\imath}_1 - \frac{\mathrm{e}^{-\mathrm{i}\nu kr}}{r}\sin\vartheta\,\tilde{\imath}_2\right].$$

From this we identify A_1 as the quantity multiplying $\tilde{\imath}_1$, and A_2 as the quantity multiplying $\tilde{\imath}_2$. Then:

$$\operatorname{curl}\tilde{A} = ikm\frac{\mathrm{e}^{-\mathrm{i}\nu kr}}{r}\left[i\nu k\sin\vartheta + \frac{\sin\vartheta}{r}\right]\tilde{\imath}_3\,.$$

Employing again the condition that $kr \gg 1$, we arrive at the expression for $\tilde{H}$, which, including the time factor, is:

$$\tilde{H} = -\,\nu mk^2\frac{\mathrm{e}^{\mathrm{i}(\omega t-\nu kr)}}{r}\sin\vartheta\,\tilde{\imath}_3\,.$$

This vector is transverse to both the direction of propagation and to the electric vector $\tilde{E}$.

Having the expressions for $\tilde{E}$ and $\tilde{H}$, the radiation energy may be calculated. As it turns out, the electric and magnetic energy densities are equal:

$$\frac{\nu^2E^2}{8\pi} = \frac{H^2}{8\pi}\,.$$

Therefore, the radiation energy can be calculated in terms of the electric vector only, as is done in this chapter.

In the original paper, Hertz retained all terms in the expressions for Φ and $\tilde{A}$, and consequently obtained values of $\tilde{E}$ and $\tilde{H}$ which are valid in all regions. He included diagrams indicating transitions in the directions as well as the magnitudes of these quantities as one goes from small to large distances (compared with λ). It is interesting to note that at small distances the vector $\tilde{E}$ is not transverse to the direction of propagation of the wave, but becomes so only in the limit of very large distances.

References p. 212

(ii) Guinier's Remarks Concerning the Radius of Gyration of a Colloidal Particle

A dilute colloidal solution exhibits a scattering of radiation whose magnitude decreases quadratically with angle near the zero angle of observation. That is, near $\vartheta = 0$, it is found empirically that $I/I_0 = 1 - p\vartheta^2$. Guinier showed that the parameter p is a measure of the radius of gyration of the particle.

The identification is made by comparing the empirical formula with a theoretically derived expression for the ratio I/I_0. The theoretical expression, for small values of ksr, is obtained by means of the expansion of the amplitude integral. We know that:

$$A \propto \int e^{i k\tilde{s} \cdot \tilde{r}} \, d\tau ,$$

where $d\tau$ is the element of volume. In expanded form:

$$A \propto \int \left[1 + ik\tilde{s} \cdot \tilde{r} - \frac{k^2}{2} (\tilde{s} \cdot \tilde{r})^2 + \ldots \right] d\tau .$$

Since $\tilde{s} \cdot \tilde{r}$ can be written as $s_x x + s_y y + s_z z$, this expression contains three integrals of the form $\int x d\tau$. If we agree to put the center from which r is measured at the center of gravity of the particle, then each of these three integrals vanishes, and accordingly so does $\int \tilde{s} \cdot \tilde{r} \, d\tau$. If we restrict ourselves to the limiting case of small angles, then it is sufficiently accurate to retain the first non-vanishing term in $\tilde{s}$. Thus:

$$A \propto \int \left[1 - \frac{k^2}{2} (\tilde{s} \cdot \tilde{r})^2 \right] d\tau = V \left[1 - \frac{k^2}{2} \int \frac{(\tilde{s} \cdot \tilde{r})^2 \, d\tau}{V} \right].$$

In these terms the intensity is:

$$I = A^2 \propto V^2 \left[1 - k^2 \frac{\int (\tilde{s} \cdot \tilde{r})^2 \, d\tau}{V} \right].$$

So far we have considered one definite orientation of the particle with respect to $\tilde{s}$. Observed intensity is really an average of the intensities produced by the particle in all orientations with respect to $\tilde{s}$. This involves the average, $<\int(\tilde{s}\cdot\tilde{r})^2\,\mathrm{d}\tau>$. A mathematically equivalent representation of this average can be made by considering the particle to be fixed while $\tilde{s}$ covers the entire solid angle 4π; this is $\int<(\tilde{s}\cdot\tilde{r})^2>\mathrm{d}\tau$. Here it is best to expand the product $(\tilde{s}\cdot\tilde{r})^2$ as $(s_x x + s_y y + s_z z)^2$. Consider one point with a particular value of x. Let $\tilde{s}$ vary over the entire solid angle and let us ask for the average $<s_x{}^2x^2>$ (see Fig. 51). This is:

$$< s_x{}^2x^2 > = \frac{x^2}{4\pi}\int_0^\pi (s\cos\vartheta)^2\cdot 2\pi\sin\vartheta\,\mathrm{d}\,\vartheta = \frac{s^2x^2}{3}\,.$$

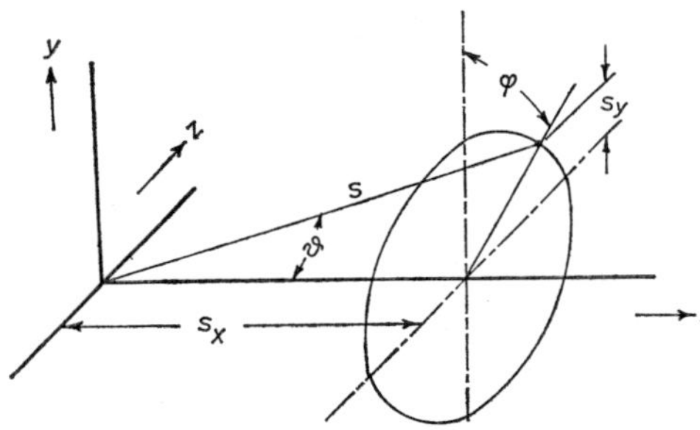

Fig. 51

The sum of this plus the corresponding terms in y and z give the desired average quantity, for cross terms vanish, as for example:

$$< s_x s_y xy > = \frac{xy}{4\pi}\int_0^{2\pi}\int_0^{\pi}(s\cos\vartheta)(s\sin\vartheta\cos\varphi)\sin\vartheta\,\mathrm{d}\vartheta\,\mathrm{d}\varphi = 0\,.$$

Thus:

$$<(\tilde{s}\cdot\tilde{r})^2> = \frac{s^2}{3}(x^2+y^2+z^2) = \frac{s^2r^2}{3}\,.$$

Finally:

$$I \propto V^2\left[1 - \frac{k^2s^2}{3}\,\frac{\int r^2\,\mathrm{d}\tau}{V}\right].$$

The term

$$\frac{\int r^2\,\mathrm{d}\tau}{V}$$

References p. 212

in this equation is equivalent to the expression for the square of the radius of gyration for our particle. Radius of gyration, r_g, is defined according to the relation $r_g^2 = \int r^2 \, \mathrm{d}M/M$, where M is total mass of the particle and the distance r is measured from the particle's center of gravity, which is taken to be the origin of coordinates. When the particle has uniform density, the ratio $\mathrm{d}M/M$ equals $\mathrm{d}\tau/V$. Therefore our equation for I becomes:

$$I \propto V^2 \left[1 - \frac{k^2 s^2}{3} r_g^2 \right] .$$

For small angles, where s, which equals

$$2 \sin \frac{\vartheta}{2} ,$$

is approximately ϑ, the ratio I/I_0 equals

$$1 - \frac{k^2}{3} r_g^2 \, \vartheta^2 .$$

Hence the empirical parameter p is identified as

$$\frac{k^2}{3} r_g^2 .$$

(*iii*) *The Exponential Correlation Function*

The correlation function, C, was assumed, in this chapter, to be the exponential $e^{-r/a}$, where r is a hypothetical measuring stick length and a is a length characteristic of the system. However, this exponential form for C can be derived [10].

We begin by focussing attention on one of the four probabilities (discussed in the chapter) relative to density fluctuations encountered at ends A and B of the stick of length r. We choose P_{11} (the probability that if end A is in material, end B will be in material), and ask how this will change when we increase the length of our measuring stick from r to a slightly larger value, $r + h$. Let end A of the stick be in material, fixed in position. The probability P_{11} will be changed if end B is close enough to a surface element so that an increase in the stick length from r to $r + h$ will cause end B to cross this surface and enter a hole (see Fig. 52). If α denotes the angle which the direction r makes with the normal to the surface, the surface will be crossed only by those sticks whose ends B lie within a sheet, parallel to the surface, of thickness $h \cos \alpha$. Considering

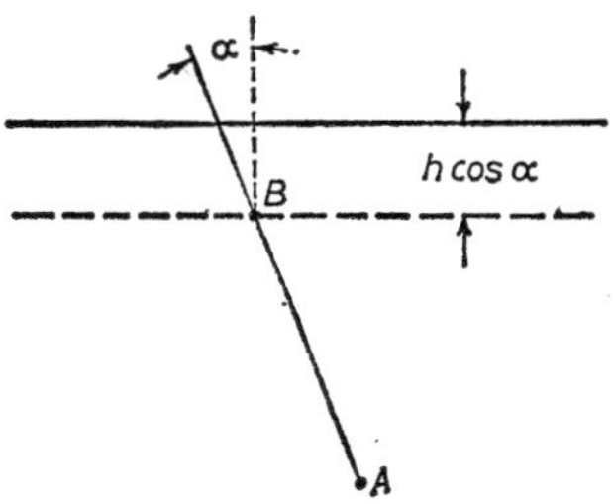

Fig. 52

all possible directions in space which could affect the probability (or, equivalently, variations in α from 0 to $\pi/2$), the average sheet thickness which has to be considered is:

$$\frac{h}{4\pi} \int_0^{\pi/2} 2\pi \cos \alpha \sin \alpha \, d\alpha = \frac{h}{4} .$$

Hence the fractional change in P_{11} due to this means of crossing the surface, will be the ratio of the average volume that end B must lie within to effect such a penetration of the surface, namely $Sh/4$, to the total material volume, V_M. In other words, we have so far the relation:

$$\delta P_{11} = - \frac{S}{V_M} \frac{h}{4} P_{11} .$$

(The negative sign is used because the change is a decrease.) On the other hand, if we consider the other side of the surface, we see that those measuring sticks which have their ends B within a hole but are near enough to cross the surface when their length is increased to $r + h$, will

References p. 212

give a positive contribution to P_{11}. This contribution will be:

$$\delta P_{11} = + \frac{S}{V_H}\frac{h}{4} P_{10}\,,$$

in which V_H is the total volume occupied by the holes. Adding up both changes in P_{11}, and replacing h in the limiting case of small values by the differential dr, we obtain the following differential equation:

$$\frac{\mathrm{d}P_{11}}{\mathrm{d}r} = -\frac{S}{4V_M} P_{11} + \frac{S}{4V_H} P_{10}\,.$$

Now, as we have shown in this chapter, $V_M = (1 - \varphi)V$, and $V_H = \varphi V$. Also, we found that both P_{11} and P_{10} can be expressed in terms of the correlation function: $P_{11} = 1 - \varphi(1 - \mathrm{C})$, and $P_{10} = \varphi(1 - \mathrm{C})$. The substitution of these values into our differential equation leads to the final result:

$$\frac{\mathrm{dC}}{\mathrm{d}r} = -\frac{S/V}{4\varphi(1 - \varphi)}\,\mathrm{C}\,.$$

For the limiting case of zero stick length, the definition of correlation function requires that C have the value of unity. In other words, our equation must meet the condition that $\mathrm{C} = 1$ for $r = 0$. Upon integrating our equation and using this boundary condition to evaluate the constant we obtain the exponential form $\mathrm{C} = \mathrm{e}^{-r/a}$, where a is given by the equation:

$$\frac{S}{V} = \frac{4\varphi(1 - \varphi)}{a}\,.$$

This is the relationship derived in the chapter through considerations of P_{diss}.

REFERENCES

1. HERTZ, H., *Ann. Physik*, 36, (1888) 1.
2. EINSTEIN, A., *Ann. Physik*, 33, (1910) 1275.
   SMOLUCHOWSKI, M., *Ann. Physik*, 25, (1908) 205.

3. Bragg, W. H., and Bragg, W. L., *X-Rays and Crystal Structure*, G. Bell & Sons, London, 1916.
4. Mie, G., *Ann. Physik*, 25, (1908) 377.
5. Guinier, A. and Fournet, G., *Small-Angle Scattering of X-Rays*, Wiley, New York, 1955.
   Guinier, A., *Ann. Physik*, 12, (1939) 161.
6. Debye, P., Anderson, H. R., Jr., and Brumberger, H., *J. Appl. Phys.*, 28, (1957) 679.
7. Brunauer, S., Emmett, P. H. and Teller, E., *J. Am. Chem. Soc.*, 60, (1938) 309.
8. Debye, P., *J. Chem. Phys.*, 31, (1959) 680.
9. Stratton, J. A., *Electromagnetic Theory*, McGraw-Hill, New York/London, 1941.
10. Debye, P., Anderson, H. R., Jr., and Brumberger, H., see Ref. 6.

CHAPTER 5

Dilute Solutions of Strong Electrolytes

The fundamental ideas that electrolytic solutions contain charged particles, and that these particles exist independently of an applied field, were introduced many years ago on the basis of early experiments in conductivity. Later work showed that factors such as degree of ionization and interactions of charged particles must be considered in formulating a conductivity theory. Our discussion here will be mainly concerned with dilute solutions of strong electrolytes, to which the idea of complete ionization with resultant electrical interactions is applicable, and we shall show the development of a theory for such solutions.

That ions exist independently of an applied field was first demonstrated by measurements of current and voltage. The magnitude of the current flowing through an electrolytic solution was found to be directly proportional to the impressed voltage, even down to the lowest measurable voltages; in other words, the electrical resistance of the solution and therefore the number of charged particles remained constant for the range of moderate voltages employed. If, on the contrary, the electrical resistance measured had shown unlimited increase as the applied voltage approached zero, it would have been reasonable to conclude that the field had created the charged particles within the solution. The postulate of the independent existence of ions was not widely accepted, however, until the 1880's, when Arrhenius presented the final convincing evidence[1]. At that time it was already well accepted that the Van 't Hoff law describing osmotic pressure provided a measure of the number of free particles in solution through the formula $\Pi = nkT$. Arrhenius showed that a conducting solution gives rise to an osmotic pressure greater than expected on the basis of this equation and the chemical formula of the solute, and he interpreted this behavior as a result of 'dissociation' of the solute.

Ostwald's Law for Dissociation

The first quantitative expression for dissociation was formulated by Ostwald [2]. To present a development of his equation, we begin by considering a molecule, AB, which dissociates into two parts, A and B. If we start with a concentration of n_0 molecules of AB per cubic centimeter, and if we let α represent the fraction dissociated, then $(1 - \alpha)n_0$ is the concentration of AB molecules which remains and the total concentration of existing ions is $2\alpha n_0$. Then by the reasoning of Guldberg-Waage (the familiar 'Mass Action Law') we have:

$$(1 - \alpha)n_0 \propto \text{rate of dissociation into ions;}$$

$$(\alpha n_0)^2 \propto \text{rate of association of ions.}$$

At equilibrium the rates are equal:

$$A_1(1 - \alpha)n_0 = A_2(\alpha n_0)^2 ,$$

where the A's represent proportionality constants. Finally:

$$\frac{\alpha^2}{1 - \alpha} = \frac{K}{n_0},$$

where $K = A_1/A_2$. This is Ostwald's law, expressing degree of dissociation as a function of concentration.

The Anomaly of Strong Electrolytes

For weak electrolytes Ostwald's law was found to hold quite well: from measurements of osmotic pressure at low concentrations it was possible to calculate K and predict the magnitude of osmotic pressure at other concentrations. However, in the case of strong electrolytes, the data could not be correlated with a single constant K. This peculiar result was called the 'Anomaly of Strong Electrolytes'.

For many years attempts were made to find a theoretical explanation of this anomaly. One of the theories advanced attempted to link the behavior of a strong electrolyte in solution to its behavior in the crystal lattice. X-ray crystallography had already shown that crystals such as

References p. 251

NaCl consist of ions arranged in a regular pattern, and it was thought that in solution the electrolyte might be ionized with the ions arranged as in the crystal but with larger lattice dimensions. However, experiments performed by G. N. Lewis [3] and others showed that this lattice theory predicted a functionally incorrect dependence of osmotic pressure on concentration. From a fundamental viewpoint, too, the postulate of a lattice structure seems untenable: the interionic forces holding a crystal-like structure together would be weakened because of the large separation distances (especially at low concentrations) and because of the high dielectric constant of the medium; under such circumstances, the ever-present Brownian motion of the ions and the solvent molecules would be expected to disrupt any ordered structure.

The Theory for Strong Electrolytes

Accordingly there was advanced a new theory, known as the Debye-Hückel theory[4], based on the well-known laws of electrostatics and the Maxwell-Boltzmann statistics. In this theory, it is assumed that the solute, being a strong electrolyte, is completely ionized; the most probable distribution of an ionic 'cloud' around a central ion is determined, and the energy involved in such a distribution is calculated. Then through thermodynamic reasoning the properties of a system may be predicted. A great mathematical difficulty is avoided by the assumption of a dilute solution; hence the theory is useful as a good approximation for low concentrations of electrolyte. Let us follow the development of the theory. We begin by considering a single ion as it moves through a dilute electrolytic solution. Let:

n_i^0 = average number of ions of type i per cubic centimeter;
q_i = total charge of the ith ion;
e = magnitude of the charge of one electron.

Under the influence of all the ions there exists in an element of volume an electric potential Φ. According to the ideas of the Maxwell-Boltzmann statistics, the most probable number of ions of type i to be found there, on a unit volume basis, is:

$$n_i = n_i^0 \, \mathrm{e}^{-q_i \Phi / kT} \, .$$

This expression approaches n_i^0 at large distances from the central ion.

Now an expression is required for the potential Φ in terms of charge density. We know from Gauss's law that the charge density ρ at some point is related to the potential at that point according to the relation

$$\text{div grad } \Phi = \frac{-4\pi\rho}{\varepsilon},$$

where ε is dielectric constant. The density of ions is simply related to ρ by the expression $\rho = \sum_i q_i n_i$, and so the desired expression is:

$$\text{div grad } \Phi = \frac{-4\pi}{\varepsilon} \sum_i q_i n_i^0 \mathrm{e}^{-q_i\Phi/kT}.$$

Thus the distribution of ions in space gives rise to a spatially varying electric potential, and this in turn influences the spatial distribution of ions. This non-linear equation is difficult to solve, and so we seek reasonable simplifications which will lead to a more tractable equation.

As a first step we note that if the ratio of electrical to thermal energy,

$$-\frac{q_i\,\Phi}{kT},$$

is small compared to unity, it is possible to expand the exponential in a Taylor series about the zero value of the argument, and retain only the first two terms: *i.e.*, $\mathrm{e}^x = 1 + x$ for small values of x. In the case of a univalent ion the ratio is

$$\frac{e\Phi}{kT},$$

which may be written in the form

$$\Phi \bigg/ \frac{kT}{e}.$$

The quantity

$$\frac{kT}{e},$$

which is interpreted as the 'fundamental potential', is at ordinary temperatures of the order of 25 millivolts, and so if the value of the ratio is to be much less than unity, Φ must be considerably smaller than 25 milli-

References p. 251

volts. How far apart must the ions be in order to make Φ small enough? To get an idea of the magnitude of this distance, let us take the case of two univalent ions in water solution, and calculate the separation distance for which Φ is 25 millivolts. The potential at one ion due to the presence of the second is represented by $e/\varepsilon l$, where l is the separation distance; this has a value of 25 millivolts when the distance l is 7.8 Å (considering water as a homogeneous liquid). Accordingly, the separation distance should be much larger than this if the ratio of electrical to thermal energy is to be small compared to unity. Only for very low ionic concentrations can this condition be approached; nevertheless, since the problem is difficult to solve otherwise, the Taylor expansion as described will be made, and (as discussed more fully below), the right order of magnitude should be obtained. Under these conditions the mathematical statement becomes:

$$\text{div grad } \Phi = -\frac{4\pi}{\varepsilon} \sum_i q_i n_i^0 \left(1 - \frac{q_i \Phi}{kT}\right).$$

Since electrical neutrality requires that $\sum_i q_i n_i^0 = 0$, we have:

$$\text{div grad } \Phi = \frac{4\pi}{\varepsilon kT} \sum_i n_i^0 q_i^2 \Phi \equiv \kappa^2 \Phi .$$

Here the quantity κ^2 has been defined as:

$$\kappa^2 \equiv \frac{4\pi}{\varepsilon kT} \sum_i n_i^0 q_i^2 .$$

It may be noted that κ has the dimensions of the inverse of length. It is equivalent to $1/\lambda$, where λ is some characteristic length. The ionic charge q_i is expressed conveniently as the product of the valence z_i and the electronic charge e. Then:

$$\kappa^2 = \frac{4\pi e^2}{\varepsilon kT} \sum_i n_i^0 z_i^2.$$

As this theory was being developed, it was encouraging to note the appearance of a particular combination of symbols in the expression for the parameter κ^2. The quantity under the summation sign had been introduced previously in an empirical manner by G. N. Lewis[5], who found this expression to be of fundamental importance in the characterization of ionic solutions. It was called the 'ionic strength'.

The development of the theory so far has been quite general; in order to apply our equation to an ionic solution, we must express div grad Φ in terms of a suitable coordinate system. By reason of symmetry the ionic distribution around some central ion should be radial, *i.e.*:

$$\text{div grad}\ \Phi = \frac{1}{r^2}\frac{\mathrm{d}}{\mathrm{d}r}\left(r^2\frac{\mathrm{d}\Phi}{\mathrm{d}r}\right) = \kappa^2\Phi\,.$$

The solution proceeds readily if we let Φ equal μ/r; then in terms of μ the equation is:

$$\frac{\mathrm{d}^2\mu}{\mathrm{d}r^2} = \kappa^2\mu\,.$$

Integration gives (in terms of Φ):

$$\Phi = \frac{A\mathrm{e}^{-\varkappa r}}{r} + \frac{B\mathrm{e}^{\varkappa r}}{r}\,.$$

The constant B must be zero because the potential must vanish as r increases without limit. Hence:

$$\Phi = \frac{A\mathrm{e}^{-\varkappa r}}{r}\,,$$

with the constant A to be evaluated.

The physical significance of the constant κ is made apparent by this relationship. If κ were zero we would have the potential due to the single central charge. The larger κ, the faster the potential decreases with distance from the central ion: this is the result of a denser, more closely packed oppositely charged ionic atmosphere. The parameter λ, the inverse of κ, represents a characteristic length, for it is a measure of the depth or persistence of the ionic atmosphere.

It should be noted that the theory predicts that at sufficiently low concentration the extent of the ionic atmosphere is determined by concentration and ionic charge and not at all by the chemical nature of

References p. 251

the ions. For example, such chemically different species as $MgCl_2$ and Na_2SO_4 would produce the same type of distribution in solution, and this is in fact found experimentally. This point will be discussed more fully later.

The constant A is evaluated through the use of Gauss's law, in the form which states that the integral of the electric displacement, εE, over the surface of a sphere, is 4π times the enclosed charge. By definition, field strength E is gradient of potential, $-\,\mathrm{d}\Phi/\mathrm{d}r$, and at the surface of an ion, considered as a sphere of radius a, this is:

$$E = A(1+\kappa a)\,\frac{\mathrm{e}^{-\varkappa a}}{a^2}\,.$$

Using Gauss's law, we obtain for the central ion of charge q_j:

$$4\pi q_j = \int_{r=a} \varepsilon E\,\mathrm{d}\,S = 4\pi a^2 \varepsilon A(1+\kappa a)\,\frac{\mathrm{e}^{-\varkappa a}}{a^2}\,.$$

In these terms the constant is:

$$A = \frac{q_j}{\varepsilon}\,\frac{\mathrm{e}^{\varkappa a}}{1+\kappa a}\,.$$

Using this evaluation of A in the expression for Φ, we may write for the electric potential at the surface of the ion:

$$\Phi_a = \frac{q_j}{\varepsilon}\cdot\frac{1}{a}\cdot\frac{1}{1+\kappa a}\,.$$

In the case of a sufficiently dilute solution, the term κa is so small compared to unity, that a satisfactory approximation is:

$$\frac{1}{1+\kappa a} = 1-\kappa a\,.$$

Hence:

$$\Phi_a = \frac{q_j}{\varepsilon a} - \frac{q_j}{\varepsilon \lambda}\,.$$

Potential at the surface of an ion is seen to be the sum of two terms: the first represents the potential due to the central charge, and the second represents a potential, arising from a virtual charge, whose magnitude depends on the thickness of the ionic atmosphere.

Assumptions of the Theory for Strong Electrolytes

Let us now examine in more detail the implications of some of the assumptions made in the development of this theory.

One of these assumptions is that the macroscopic rather than microscopic value for dielectric constant can be used. The reasonableness of this approximation can be discussed in terms of the parameter κ, which is a measure of the depth of ionic atmosphere. A calculation indicates that for the dilute solutions that are considered, the ionic atmospheres are extensive enough to include sufficient solvent molecules to make valid the use of the macroscopic ε. Let us consider a specific example. Suppose we have a solution of sodium chloride in water, of concentration 0.001 molar. This represents about 12×10^{17} ions per cubic centimeter. Using (for order of magnitude) the macroscopic dielectric constant of water, $\varepsilon = 80$, we calculate a κ equal to 10^6 cm^{-1}. The thickness of the ionic atmosphere is measured by the inverse of κ, and is calculated to be $\kappa^{-1} = \lambda = 100$ Å. An atmosphere of this great an extent would be expected to include many water molecules, and so it seems reasonable to use the macroscopic dielectric constant.

It was also assumed that concentration of ions is given by the most probable value — fluctuations in the ionic atmosphere were neglected. A more accurate calculation including fluctuations has been made[6], but the result is not fundamentally different.

In forming an expression for div grad Φ, the assumption was made that $e^{-e_i\Phi/kT}$ is approximated by $1 - (e_i\Phi/kT)$, thus enabling us to proceed to a solution. Let us consider a one-dimensional case, for which an exact solution exists, in order to see how much error is involved in making the approximation. Suppose we have a solution of uni-univalent ions contained between parallel plate electrodes whose area is so large that edge effects can be neglected: if we construct the x coordinate axis normal to the faces, the macroscopic field strength depends only on x. When a difference of potential is applied between the plates, the oppositely charged ions cluster about each plate. There is a critical voltage below

References p. 251

which there will be no discharge of ions (provided the plates have not been coated with platinum black), and we shall consider that we have a voltage below this value. Since there are two ions of equal charge, the fundamental differential equation for potential reduces to a simpler, though non-linear form:

$$\operatorname{div}\operatorname{grad}\Phi = \frac{4\pi}{\varepsilon}\left(en_0 \mathrm{e}^{-\frac{e\Phi}{kT}} - en_0 \mathrm{e}^{\frac{e\Phi}{kT}}\right) = \frac{8\pi n_0 e}{\varepsilon}\sinh\frac{e\Phi}{kT}.$$

Here n_0 represents the ionic concentration, which is the same for both ions. It should be clearly understood that this simplification is possible only because ions of exactly opposite charge are being considered. Since in addition we are dealing with a potential which depends only on the x coordinate, the equation reduces to a total non-linear one:

$$\frac{\mathrm{d}^2\Phi}{\mathrm{d}x^2} = \frac{8\pi n_0 e}{\varepsilon}\sinh\frac{e\Phi}{kT}\dagger.$$

A dimensionless equation may be obtained through the introduction of some new units. If we make use of the 'fundamental potential' defined as $\Phi_0 = kT/e$, and let $\varphi = \Phi/\Phi_0$, we obtain:

$$\frac{\mathrm{d}^2\varphi}{\mathrm{d}x^2} = \left(\frac{8\pi n_0 e^2}{\varepsilon kT}\right)\sinh\varphi\,.$$

The term in parentheses is the quantity κ^2, and so because κ is the inverse of λ, we are measuring length in terms of the thickness of the ionic atmosphere. The equation can be simplified and made more meaningful by writing it in terms of the quantity ξ, which we define as $\xi \equiv \kappa x = x/\lambda$. Thus:

$$\frac{\mathrm{d}^2\varphi}{\mathrm{d}\xi^2} = \sinh\varphi\,.$$

† Because we are assuming low concentration, we do not consider the effect due to the finite ionic volume. At high concentrations a certain amount of mechanical work would be needed to pack the bulky hydrated ions into the space near the electrodes. This would have to be included in the Boltzmann expression and the problem would be more involved.

Integration now proceeds smoothly. Multiply both sides by

$$\frac{d\varphi}{d\xi}.$$

Then:

$$\frac{d^2\varphi}{d\xi^2}\cdot\frac{d\varphi}{d\xi} = \frac{d}{d\xi}\left[\frac{1}{2}\left(\frac{d\varphi}{d\xi}\right)^2\right] = \sinh\varphi\,\frac{d\varphi}{d\xi} = \frac{d}{d\xi}(\cosh\varphi + \text{constant})\,.$$

Integration gives:

$$\frac{1}{2}\left(\frac{d\varphi}{d\xi}\right)^2 = \cosh\varphi + \text{constant}\,.$$

In order to evaluate the constant of integration, let us consider the problem from the standpoint of the dimension ξ. The electrode spacing is very large compared with λ, and so we expect that at distances far removed from the electrode the potential will be changing only very slowly. In terms of the dimension ξ this change should be very small indeed, and so (considering only one electrode) we may put for one boundary condition:

$$\lim_{\xi\to\infty}\frac{d\varphi}{d\xi} = 0\,.$$

A second boundary condition is that the potential (again considering only one electrode) approaches 0 as ξ increases:

$$\lim_{\xi\to\infty}\varphi = 0\,.$$

The constant of integration must, then, have the value -1.

The second integration proceeds through a simple identity:

$$\frac{1}{2}\left(\frac{d\varphi}{d\xi}\right)^2 = \cosh\varphi - 1 = \frac{e^{\varphi} + e^{-\varphi} - 2}{2} = \frac{(e^{\varphi/2} - e^{-\varphi/2})^2}{2}\,;$$

or:

$$\frac{d\varphi}{d\xi} = -\,(e^{\varphi/2} - e^{-\varphi/2})\,^*; \qquad \text{or} \int\frac{d\varphi}{e^{\varphi/2} - e^{-\varphi/2}} = -\int d\xi = \xi_0 - \xi\,,$$

* Our problem requires the negative square root, since the derivative must be negative.

References p. 251

where ξ_0 is the second constant of integration. With the substitution $e^{\varphi/2} = t$, our equation becomes:

$$\int \frac{2dt}{t^2 - 1} = \ln \frac{t-1}{t+1} = \xi_0 - \xi \, .$$

Taking antilogarithms we obtain:

$$\frac{e^{\varphi/2} - 1}{e^{\varphi/2} + 1} = e^{\xi_0} \, e^{-\xi} \, .$$

To evaluate the integration constant ξ_0, we make use of the fact that at the surface of the electrode ξ equals 0. There the potential is some given value $\varphi = eV/kT \equiv v$. Hence:

$$e^{\xi_0} = \frac{e^{v/2} - 1}{e^{v/2} + 1} \, .$$

The final expression with all constants evaluated is:

$$\frac{\varphi}{2} = \ln \frac{(e^{v/2} + 1) + (e^{v/2} - 1)e^{-\xi}}{(e^{v/2} + 1) - (e^{v/2} - 1)e^{-\xi}} \dagger .$$

If the potential at the electrode is small the expression above reduces to the exponential form $\varphi = ve^{-\xi}$ (which is the form of the expression for potential obtained with the Debye-Hückel theory). The rate of change of potential at the electrode for small values of v is, then:

$$\left(\frac{d\varphi}{d\xi}\right)_{\xi=0} = -\, v \, .$$

The more accurate expression $d\varphi/d\xi = e^{-\varphi/2} - e^{\varphi/2}$ must be referred to for large values of v, and shows that the potential near the electrode surface changes at the rate:

$$\frac{d\varphi}{d\xi} = e^{-v/2} - e^{v/2} \, .$$

† The presence of ionic atmosphere also greatly influences electrical capacitance of such a condenser. For details see the Appendix, pp. 248, 249.

This is a more rapid rate of change with distance than is given by the previous simple expression; to a first order approximation, however, it is identical.

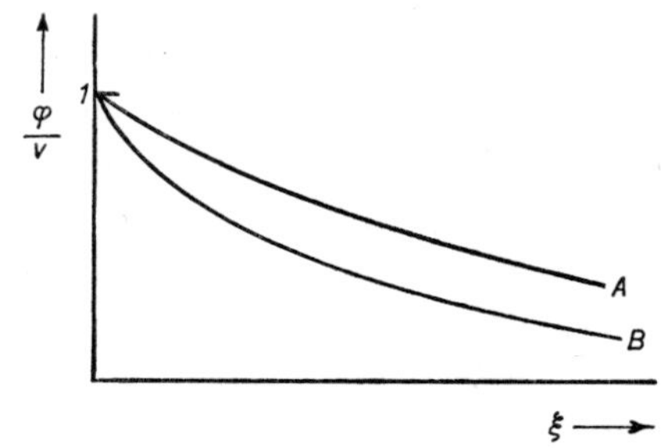

Fig. 53

The graph (see Fig. 53) summarizes the results. Curve A shows how φ varies when the applied voltage is small, and curve B represents the situation for large potential. The interpretation is that when large potential differences are applied between the electrodes, ions concentrate near each electrode to a greater extent than is indicated by the approximation. This means that in the case of large applied potential difference, significant error may be introduced when the calculation follows the simpler line of reasoning. Nevertheless, even in this case all the important phenomena are represented in the simple equation, and it should serve as a good first approximation.

The solution for small potential to this one-dimensional problem corresponds to the solution (for the three dimensional case) presented by the Debye-Hückel theory. This correspondence indicates that the Taylor approximation used in the Debye-Hückel theory may introduce significant error when ionic concentration is high, but that in any case the theory provides a good first approximation.

Evaluation of Theory by Experiment: Debye-Hückel Theory Compared with Classical Dissociation Theory

The next step in the consideration of the Debye-Hückel theory for strong electrolytes is to evaluate its reliability as compared with that of classical dissociation theory when applied to such solutions. To do this, we may

References p. 251

compare theoretical calculations with a measureable property, the dependence of osmotic pressure on concentration.

Osmotic Pressure as Predicted by Dissociation Theory

As a first step, let us consider the classical theory. It begins with the Van 't Hoff law, $\Pi = nkT$, and assumes that the concentration of particles in solution, n, is the result of an equilibrium, expressed in terms of an equilibrium constant, K, between ions and neutral molecules. The concentration of particles in solution, n, is less than the concentration of ions, n_T, that would result if dissociation were complete. If Van 't Hoff's law is expressed in the form $\Pi = (1 - f)n_T kT$, then f may be interpreted as the degree of incomplete dissociation.

An experimental value for f can be obtained from measurements of osmotic pressure; the graph in Fig. 54 gives a rough picture of such

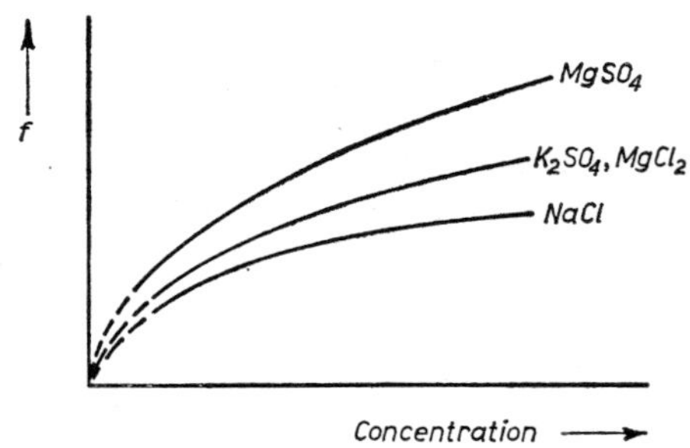

Fig. 54

values of f as a function of concentration, for various electrolytes. Values of f are seen to increase with concentration, and this is reasonable in terms of dissociation theory. However, several discrepancies between theory and experiment are apparent. (1) It is found experimentally that the slope at the origin tends toward infinity like the square root of concentration, for all ionic species. But the dissociation theory predicts that near zero concentration the function f should have a finite tangent whose value depends on the parameters of the given system*. (2) The values of f are seen to be higher for K_2SO_4 than for NaCl, and this behavior is difficult to interpret from the standpoint of dissociation theory. The formation of a molecule of K_2SO_4 would require a three-

* See the Appendix, pp. 249, 250.

body collision; according to statistics, such a collision should occur less frequently than a two-body collision, and so K_2SO_4 should be more highly dissociated than NaCl. (3) Graphs for the chemically different species K_2SO_4 and $MgCl_2$ are experimentally very much the same. A basic idea of the theory, however, is that chemically different substances would have different degrees of dissociation.

Osmotic Pressure and the Theory for Strong Electrolytes

Now let us see what the theory for strong electrolytes predicts for the dependence of osmotic pressure on concentration. Consider a cylinder which contains a dilute solution of electrolyte, and which is fitted with a semi-permeable piston. When the piston is pushed down the solution becomes more concentrated and a certain amount of work is done; if the piston is depressed in a reversible and isothermal manner, the amount of work done depends only on the initial and final states and not on the particular path:

$$W_T = -\int_{V_{\text{in}}}^{V_f} \Pi \, \mathrm{d}V ,$$

where W_T is the total amount of work and Π is the osmotic pressure. Now we may consider this operation to be accomplished in another way. Suppose that first the ions are discharged, then the piston is moved down (reversibly and isothermally), and then the charges are replaced. Since the initial and final states are the same as before, an equal amount of work is done and we may write:

$$-\int_{V_{\text{in}}}^{V_f} \Pi \, \mathrm{d}V = -\int_{V_{\text{in}}}^{V_f} \Pi_{\text{ideal}} \, \mathrm{d}V + W ,$$

where Π_{ideal} represents the osmotic pressure which would be developed by the dilute solution if the solute particles were uncharged, and W is the net electrical work. Differentiation with respect to the final volume gives:

$$\Pi = \Pi_{\text{ideal}} - \frac{\partial W}{\partial V} .$$

As will be shown, the sign of the term $\partial W/\partial V$ is positive, so that real pressure is less than the ideal pressure.

The electrical work, W, can be calculated from an expression obtained

References p. 251

from our theory for strong electrolytes. We derived an expression for electrostatic potential at the surface of the central ion:

$$\Phi_a = \frac{q_i}{\varepsilon a} - \frac{q_i}{\varepsilon} \kappa \, .$$

Here we need be concerned only with the second term, since the first is independent of volume and therefore would not effect an expression involving net work. If we suppose that each ion of type i is charged to some fraction η of its full charge, then the ionic charge is ηq, and the magnitude of the atmosphere around it is indicated by the product $\eta\kappa$. Hence the expression for the volume-dependent part of potential at any degree of charge is:

$$\Phi_a = - \frac{q_i}{\varepsilon} \kappa \eta^2 \, .$$

The electrical work, which includes both the charging of the ion and the building up of the atmosphere, is obtained by integration:

$$W_i = \int_{\text{in}}^{f} \Phi_a \, \mathrm{d}q_i \, .$$

The element of charge $\mathrm{d}q_i$, is $q_i \, \mathrm{d}\eta$. The final expression which includes the effect of all the ions is:

$$W = \sum_i W_i = \sum_i N_i \frac{q_i^2}{\varepsilon} \kappa \int_0^1 \eta^2 \, \mathrm{d}\eta = - \sum_i N_i \frac{q_i^2}{3\varepsilon} \kappa \, .$$

It should be noted that we are assuming that the dielectric constant remains the same between initial and final states.

The expression $\partial W / \partial V$ may now be evaluated. Volume appears in the expression for W through the quantity κ, which is defined:

$$\kappa^2 = \frac{4\pi}{\varepsilon k T} \sum_i \frac{N_i q_i^2}{V} \, .$$

Since:

$$\frac{\partial W}{\partial V} = \frac{\mathrm{d}W}{\mathrm{d}\kappa} \frac{\partial \kappa}{\partial V} ,$$

then:

$$\frac{\partial W}{\partial V} = \sum_i \frac{N_i q_i^2}{6\varepsilon} \frac{\kappa}{V} = \sum_i n_i^0 \frac{q_i^2}{6\varepsilon} \kappa \, .$$

This is always positive.

The total osmotic pressure developed in the solution we have expressed as:

$$\Pi = \Pi_{\text{ideal}} - \frac{\partial W}{\partial V} \, .$$

This may now be written:

$$\Pi = \sum_i n_i^0 kT - \sum_i n_i^0 \frac{q_i^2}{6\varepsilon} \kappa = \sum_i n_i^0 kT \left[1 - \frac{1}{6} \frac{q_i^2 \varepsilon \lambda}{kT} \right] .$$

The term

$$\frac{1}{6} \frac{q_i^2 \varepsilon \lambda}{kT}$$

may be regarded as an expression of deviation from ideality. It is the exact counterpart of the quantity we termed f in our discussion of dissociation theory; but unlike f, which was interpreted as the degree of incomplete dissociation, this term arises from the concept of long-range electrical forces acting between the ions of completely dissociated molecules *. Because of the presence of λ in the term, we may conclude that the theory for strong electrolytes predicts that deviations from ideal osmotic pressure should vary with the square root of concentration. This prediction, unlike that of the classical theory, is in agreement with experimental measurements.

In other ways, too, the theory of strong electrolytes is a better expression of experimental dependencies. According to the equation, deviation from ideal osmotic pressure should show no dependence on the chemical nature of the dissolved substance but should increase with number of particles and the ionic charge. Reference to the graph of experimental measurements shows that these predictions are valid.

Thermodynamic Properties Calculated from Debye-Hückel Theory

The theory for strong electrolytes provides the means to calculate other

* For a discussion of the interaction energy requirement for a uniform particle distribution, see the Appendix, pp. 250, 251.

thermodynamic properties of non-ideal systems: the expression for electrical work involved in building the ionic atmosphere can be incorporated into the pertinent thermodynamic expressions. This may be done directly (as in the calculation of osmotic pressure we have just made) by adding a separate work term in the appropriate equation. But it is convenient to calculate, instead, one of several fundamental thermodynamic quantities which incorporate the electrical work term, and which can be used in place of their 'ideal' counterparts. One of these quantities is chemical potential. And from this we can calculate the quantity called activity, which can be substituted for concentration into any thermodynamic equation already derived for ideal solution. Thus the vast body of thermodynamic equations describing the properties of ideal solution is immediately applicable to the description of properties of dilute solutions of strong electrolytes.

Chemical Potential

Let us begin the derivation of chemical potential by considering the pertinent thermodynamic relationships. According to the first and second laws of thermodynamics:

$$\mathrm{d}U = T\mathrm{d}S - P\mathrm{d}V + \delta W\,,$$

where δW is the work other than pressure-volume work. If we subtract $\mathrm{d}(TS)$ and add $\mathrm{d}(PV)$ to both sides of the equation, we obtain:

$$\mathrm{d}(U + PV - TS) = -\,S\mathrm{d}T + V\mathrm{d}P + \delta W\,.$$

The quantity $U + PV - TS$ is termed Ψ, the Gibbs free energy. At constant temperature and pressure:

$$\mathrm{d}\Psi = \delta W\,.$$

The work in our case is the electrical work that accompanies the charging of ions and the building of an atmosphere; in other words, it is the work that brings the system from the ideal to the real state. Thus in integrated form, the above equation becomes:

$$\Psi_{\mathrm{real}} - \Psi_{\mathrm{ideal}} = W\,.$$

Now Ψ_{real} is a function of the concentration of components of the system. If temperature and pressure are constant we may write:

$$\mathrm{d}\Psi_{\text{real}} = \omega_1 \mathrm{d}N_1 + \omega_2 \mathrm{d}N_2 + \ldots, \quad \text{where } \omega_i = \left(\frac{\partial \Psi_{\text{real}}}{\partial N_i}\right)_{T,\,P,\,N_j}.$$

We can evaluate these ω's, which are the chemical potentials of the components, by referring to the relationships derived above:

$$\Psi_{\text{real}} = \Psi_{\text{ideal}} + W.$$

The work W is obtained by means of the Debye-Hückel theory; we must now evaluate Ψ_{ideal} in terms of concentration.

The quantity Ψ_{ideal} for a mixture is simply the sum of the free energies of the individual pure components, plus the free energy of mixing:

$$\Psi_{\text{ideal}} = \sum_i N_i \psi_i + \Delta\Psi_{\text{mixing}}.$$

It will be shown that $\Delta\Psi_{\text{mixing}}$ depends only on temperature and composition, and so is a constant for a given mixture at constant temperature: energy from an external source such as an electric, magnetic or gravitational field would not change the magnitude of this term – it would only give rise to a new set of values for the individual free energies, ψ_i.

We can evaluate the free energy of mixing by considering a hypothetical cyclic process. Suppose that the solution has been evaporated at constant temperature, and that the gases have been separated and maintained at the partial pressures which are the equilibrium pressures of the vapor over the liquid mixture. This process involves no free energy change. Suppose that each gas is then compressed to the vapor pressure, p^0, that exists over the pure material. In this step, there is a free energy change due to compression. However, no free energy change results from the ensuing condensation of each gas at its own vapor pressure. A cycle is completed by mixing the components, and since total free energy change must be zero, we can write:

$$\Delta\Psi_{\text{mixing}} + \Delta\Psi_{\text{compression}} = 0.$$

Let us evaluate the free energy of compression of the gases. By definition, we know:

$$\Delta\Psi = \Delta U + \Delta(PV) - \Delta(TS).$$

References p. 251

For ideal gases at constant temperature the expression for free energy of compression reduces to:

$$\Delta\Psi = - T\Delta S .$$

The entropy change is obtained from the combined first and second laws of thermodynamics:

$$\Delta S = \int \mathrm{d}S = \int \frac{\mathrm{d}U + P\mathrm{d}V}{T} .$$

During compression of a perfect gas at constant temperature the internal energy is constant, so that:

$$\Delta S = \int \frac{P}{T} \mathrm{d}V .$$

For one mole of a perfect gas, $PV = RT$, so that:

$$\Delta S = \int R \frac{\mathrm{d}V}{V} = R \ln \frac{V^0}{V} = R \ln \frac{p}{p^0} .$$

The fraction p/p^0 due to a given component i in an ideal liquid mixture is equal to the mole fraction, γ_i, of that component. Hence for one molecule:

$$\Delta s_i = k \ln \gamma_i .$$

Taking into account the total number of molecules of all components, the total entropy of compression is:

$$\Delta S = \sum_i N_i \Delta s_i = \sum_i N_i k \ln \gamma_i .$$

This evaluation of the entropy of compression may now be used in calculating the free energy of mixing, for we have shown by means of our hypothetical cyclic process that:

$$\Delta\Psi_{\text{mixing}} = - \Delta\Psi_{\text{compression}} = T\Delta S .$$

Hence:

$$\Delta\Psi_{\text{mixing}} = \sum_i N_i kT \ln \gamma_i .$$

The two terms in the expression for Ψ_{ideal} are now combined:

$$\Psi_{\text{ideal}} = \sum_i N_i(\psi_i + kT \ln \gamma_i) .$$

The next step in the evaluation of Ψ_{real} is to add in the work term. From the theory for strong electrolytes we obtain the expression:

$$W = - \sum_i N_i \frac{q_i^2 \kappa}{3\varepsilon} .$$

It will be recalled that the quantity κ is a function of the square root of concentration; ε is dielectric constant. The expression for Ψ_{real} is:

$$\Psi_{\text{real}} = \sum_i N_i \left(\psi_i + kT \ln \gamma_i - \frac{q_i^2 \kappa}{3\varepsilon} \right) .$$

We are now in a position to put this equation into the differential form in which chemical potentials appear. In performing this operation we must remember that concentration is involved in the quantities N_i, γ_i, and κ. Thus, by use of the 'product rule' $[\mathrm{d}(xy) = x\,\mathrm{d}x + y\,\mathrm{d}x]$, we write:

$$\mathrm{d}\Psi_{\text{real}} = \sum_i \mathrm{d}N_i \left(\psi_i + kT \ln \gamma_i - \frac{q_i^2 \kappa}{3\varepsilon} \right)$$

$$+ \sum_i N_i \left(\frac{kT}{\gamma_i} \mathrm{d}\gamma_i - \frac{q_i^2}{3\varepsilon} \sum_j \frac{\partial \kappa}{\partial N_j} \mathrm{d}N_j \right) .$$

It should be noted that the subscript j is used to indicate that this summation must be performed independently of the summation over i; however, both i and j refer to the same series of terms in N.

Let us consider, in the above equation, the term

$$\sum_i N_i \frac{\mathrm{d}\gamma_i}{\gamma_i} .$$

By definition,

$$\gamma_i = \frac{N_i}{\sum_j N_j} .$$

References p. 251

Using the 'quotient rule' of calculus, we find:

$$\mathrm{d}\gamma_i = \frac{\mathrm{d}N_i \sum_j N_j - N_i \sum_j \mathrm{d}N_j}{(\sum_j N_j)^2}\,.$$

Hence:

$$\sum_i N_i \frac{\mathrm{d}\gamma_i}{\gamma_i} = \sum_i \frac{\mathrm{d}N_i \sum_j N_j - N_i \sum_j \mathrm{d}N_j}{\sum_j N_j}$$

$$= \frac{\sum_i \mathrm{d}N_i \sum_j N_j - \sum_i N_i \sum_j \mathrm{d}N_j}{\sum_j N_j} \equiv 0\,.$$

The expression for Ψ_{real} accordingly reduces to:

$$\mathrm{d}\Psi_{\mathrm{real}} = \sum_i \mathrm{d}N_i \left(\psi_i + kT \ln \gamma_i - \frac{q_i^2 \kappa}{3\varepsilon} \right) - \sum_i N_i \frac{q_i^2}{3\varepsilon} \sum_j \frac{\partial \kappa}{\partial N_j} \mathrm{d}N_j\,.$$

We can write this in a more convenient form by exchanging the subscript designations in the last term so that all variations will have the subscript *i*. Thus for the final term we substitute:

$$- \sum_j N_j \frac{q_j^2}{3\varepsilon} \sum_i \frac{\partial \kappa}{\partial N_i} \mathrm{d}N_i \text{ or } - \sum_i \mathrm{d}N_i \sum_j N_j \frac{q_j^2}{3\varepsilon} \frac{\partial \kappa}{\partial N_i}\,.$$

Finally:

$$\mathrm{d}\Psi_{\mathrm{real}} = \sum_i \mathrm{d}N_i \left(\psi_i + kT \ln \gamma_i - \frac{q_i^2 \kappa}{3\varepsilon} - \sum_j N_j \frac{q_j^2}{3\varepsilon} \frac{\partial \kappa}{\partial N_i} \right)$$

$$= \sum_i \mathrm{d}N_i \cdot \omega_i(P, T, \gamma)\,.$$

By definition, the ω's are chemical potentials. So we write:

$$\omega_i = \psi_i + kT \ln \gamma_i - \frac{q_i^2 \kappa}{3\varepsilon} - \sum_j N_j \frac{q_j^2}{3\varepsilon} \frac{\partial \kappa}{\partial N_i}\,.$$

It is apparent that in this expression for chemical potential, the first two terms, $\psi_i + kT \ln \gamma_i$, represent the evaluation for the ideal case, *i.e.*, at infinite dilution. Furthermore, the logarithmic term represents the effect of mixing. The third and fourth terms are due to electrical forces and result from application of the Debye-Hückel theory.

This general expression is valid for all components of a dilute solution of ions. A simplification is possible if we calculate instead two expressions: one for the solvent, and one for dissolved species. We begin by letting N_0 be the number of uncharged solvent molecules and $N_1, N_2, \ldots$ be the number of dissolved ions. The solution is assumed to be dilute, and so the total volume may be considered as the sum of partial volumes:

$$V = N_0 v_0 + \sum_i N_i v_i \,.$$

We shall first calculate the chemical potential of the solvent molecules. The term we must evaluate is

$$\sum_j N_j \frac{q_j^2}{3\varepsilon} \frac{\partial \kappa}{\partial N_0} \,;$$

(The term

$$\frac{q_i^2 \kappa}{3\varepsilon}$$

is zero because the molecule is uncharged.) We begin with the definition

$$\kappa^2 = \frac{4\pi}{\varepsilon k T} \sum_i N_i q_i^2 / V \,.$$

To obtain

$$\frac{\partial \kappa}{\partial N_0}$$

we write:

$$2\kappa \frac{\partial \kappa}{\partial N_0} = \frac{\partial \kappa^2}{\partial N_0} = -\frac{4\pi}{\varepsilon k T} \frac{\sum_i N_i q_i^2}{V^2} v_0 = -\kappa^2 \frac{v_0}{V} \,.$$

Solving:

$$\frac{\partial \kappa}{\partial N_0} = -\frac{\kappa}{2} \frac{v_0}{V} \,.$$

The chemical potential for the solvent is, then:

$$\omega_0 = \psi_0 + kT \ln \gamma_0 + \kappa v_0 \sum_j \frac{n_j q_j^2}{6\varepsilon} \,.$$

Here n_j is concentration per unit volume, N_j/V.

References p. 251

In evaluating chemical potential of any dissolved species, it is sufficiently accurate to write

$$\frac{\partial \kappa^2}{\partial N_i} = \frac{4\pi}{\varepsilon kT} \frac{q_i^2}{V},$$

because changes in volume due to the ionic species are negligible for a dilute solution. The sum

$$-\sum_j N_j \frac{q_j^2}{3\varepsilon} \frac{\partial \kappa}{\partial N_i}$$

is in this case

$$-\frac{q_i^2 \kappa}{6\varepsilon};$$

this follows from the definition of κ^2. Finally:

$$\omega_i = \psi_i + kT \ln \gamma_i - \frac{q_i^2 \kappa}{2\varepsilon}.$$

For dissolved ions, the effect of electrical forces is seen to be due to their ionic charge. The small changes in volume which give rise to the entire electrical effect in the case of the solvent, would give rise to only a small correction term here.

Osmotic Pressure

An example of the use of our expressions for chemical potential is in the calculation of osmotic pressure of a dilute solution of strong electrolyte. (It will be recalled that this calculation was carried out above by the direct method of adding a separate work term to the 'ideal' expression.) Suppose we have pure water, at temperature T and pressure P, separated by a semi-permeable membrane from an ionic solution at temperature T and pressure $P + \Pi$, where Π is osmotic pressure. Let us focus attention on one water molecule. In the pure solvent at pressure P, the chemical potential of the molecule is ω_0. The change in chemical potential which results when this molecule passes through the membrane into the solution, is $\Delta\omega_0$, which at equilibrium equals zero. In the solution the pressure is $P + \Pi$; for convenience we define a chemical potential of the pure solvent at this pressure as ω_0'. The $\Delta\omega_0$ can be thought of as $\omega_0' - \omega_0$ plus free energy change (at constant pressure) due to change in concentration:

$$\Delta\omega_0 = 0 = \omega_0' - \omega_0 + kT \ln \gamma_0 + \kappa v_0 \sum_i \frac{n_i q_i^2}{6\varepsilon}.$$

We can evaluate the term $\omega_0' - \omega_0$, which in different notation is $\omega_0(P + \Pi) - \omega_0(P)$, by use of a simple approximation:

$$\omega_0' - \omega_0 = \omega_0(P + \Pi) - \omega_0(P) \cong \left(\frac{\partial \omega_0}{\partial P}\right)_T \Pi.$$

From the definition of chemical potential we know that its isothermal variation with pressure equals partial molecular volume:

$$\left(\frac{\partial \omega_0}{\partial P}\right)_T = v_0.$$

Collecting terms, we find:

$$\Delta\omega_0 = 0 = \Pi v_0 + kT \ln \gamma_0 + \kappa v_0 \sum_i \frac{n_i q_i^2}{6\varepsilon}.$$

This equation can be written so as to yield an expression for Π similar in form to the expression for Van 't Hoff's law. By definition:

$$\gamma_0 = \frac{N_0}{N} = 1 - \frac{\sum_i N_i}{N},$$

and because the mole fraction of water is so close to unity we may write:

$$\ln \gamma_0 = \ln \left[1 - \frac{\sum_i N_i}{N}\right] \cong - \frac{\sum_i N_i}{N}.$$

Then:

$$\Pi = kT \frac{\sum_i N_i}{N v_0} - \kappa \sum_i \frac{n_i q_i^2}{6\varepsilon} = \sum_i n_i kT \left[1 - \kappa \frac{q_i^2}{6\varepsilon kT}\right].$$

This expression is, of course, identical to the one derived earlier in this chapter by a different method of applying the theory for strong electrolytes. It is interesting to note that the bracketed expression represents

References p. 251

Bjerrum's empirical osmotic coefficient [7]. We may also point out that the term $\kappa q_i^2/6\varepsilon kT$ constitutes, aside from the factor 1/6, the ratio of two energies – the energy of two electrostatic charges of strength q_i separated by a distance λ and immersed in a medium of dielectric constant ε; and the thermal energy kT. It is reasonable to expect that this ratio should appear in the correction term, because it is of fundamental importance in the statistical theory used to describe ionic atmosphere.

Activity

Our expression for chemical potential can be used to obtain a theoretical expression for another useful thermodynamic quantity, activity, which has long been in use as an empirical quantity calculated from experimentally determined colligative properties. In the expression for ω_i, the last three terms, which involve concentration, may be combined into one term so that the equation is of the form:

$$\omega_i = \psi_i + kT \ln a \,,$$

which is similar to the equation for ideal solution, $\omega_i = \psi_i + kT \ln \gamma_i$. The quantity a is what we have termed activity, and is:

$$a = \gamma_i \exp\left[-\frac{1}{3\varepsilon kT}\left(q_i^2\kappa + \sum_j N_j q_j^2 \frac{\partial\kappa}{\partial N_i}\right)\right].$$

Of course, this theoretical expression is applicable only to dilute solutions of strong electrolytes because of the limitations of the theory it is based on. But as we have already pointed out, this expression for activity can be used in place of concentration in any thermodynamic equation derived for ideal solution.

Sometimes one finds activity written in terms of mole fraction, as $a_i = f_i\gamma_i$; f_i is called activity coefficient. Comparison with the equation just developed shows that f is identical to the exponential factor, which is always less than unity. For dilute ionic solutions, the activity of a species is always less than its mole fraction.

Conductivity

Electrical conductivity has long been recognized as an important measurable property of ionic solutions; in fact, as mentioned in the beginning of this chapter, experimentation in conductivity led to the idea of ionization. Yet for many years there was no conductivity theory that gave satisfactory agreement with experimental results. Let us now see how well the ideas of the Debye-Hückel theory apply to a consideration of this irreversible process.

First of all, for the purpose of discussion, we may derive a simple expression for conductivity in terms of ionic concentration. If an electric field E is applied across a conducting solution, each ion is acted on by two opposing forces: (1) the electrical force, Eq_i, and (2) the force due to friction, $\rho_i v_i$, where v_i is the average velocity of the ion and ρ_i is the friction constant. Under conditions of dynamic equilibrium the forces balance, there is no acceleration, and the ion moves with a constant average velocity of $v_i = q_i E/\rho_i$. The corresponding particle current density, which is the number of particles of type i that pass through an area of one square centimeter in one second, is $n_i v_i$ or $n_i q_i E/\rho_i$. Hence the electric current density J, which is particle current density multiplied by the charge per particle, is:

$$J = \sum_i \frac{n_i q_i^2}{\rho_i} E .$$

If we assume that E represents the macroscopic field, then we can define conductivity Λ as J/E or:

$$\Lambda = \sum_i \frac{n_i q_i^2}{\rho_i} .$$

The charge q_i is known, and if it is further assumed that the friction constant ρ_i is concentration independent, then Λ should be a linear measure of the density of ions in solution.

Experimentally it is found that conductivity does not depend linearly on concentration. The first attempts to explain this anomaly were based on dissociation theory, but such explanations were only qualitatively satisfactory; in important details, theoretical predictions were incorrect. For example, Kohlrausch [8] found that the graph of (Λ/C) *vs.* C for uni-univalent ions is not linear at low concentrations, as one might expect from dissociation theory, but rather falls off with infinite slope at zero concentration; in this limiting case a nearly linear relationship is seen in the graph of (Λ/C) *vs.* $C^{1/2}$. Kohlrausch also found that roughly the

References p. 251

same curve is obtained from conductivity measurements of all uni-univalent materials, regardless of their chemical nature, and this result is contrary to the ideas of dissociation theory. It will be recalled that similar discrepancies became apparent when dissociation theory was used to explain experimentally determined variations of osmotic pressure with concentration.

In the theory for strong electrolytes, it is assumed that there is total dissociation, but corrections are made for electrical interaction between the ions. Let us now use these ideas to develop a conductivity theory. We begin by considering again the microscopic picture of each ion. When we assume that each ion is surrounded by an oppositely charged atmosphere, it is reasonable to expect that the ion should have less mobility than it would have in a non-interacting system. We can advance two reasons for this: (1) the effective field acting on each ion is less than the macroscopic field, and (2) the medium surrounding each ion carries a net charge and so is moved in the opposite direction, thus decreasing the velocity of the ion with respect to the plates.

The magnitude of the field acting on each ion is affected by the time required for the building up or decay of the ionic atmosphere. We may look at the situation this way: if the central ion were suddenly discharged, its atmosphere would diffuse away, and the process would be characterized by some finite relaxation time. If instead of being discharged, the ion is pulled through its medium, it leaves behind it a decaying oppositely charged atmosphere while it continuously attempts to build up an atmosphere at each point. Because of the finite value of the relaxation time, the moving ion seems to be always pulling away from its atmosphere. Obviously, the longer the relaxation time and the higher the velocity of the ion, the greater will be the distortion of the surrounding cloud of opposite charge. The effect of this dissymmetry is the creation of a field oppositely directed to the macroscopic field. Hence the effective electric field acting on the ion could be written as $E - f$, where f represents the field caused by the dissymmetry. A calculation for f has been made, and can be found in the literature [9]. It is based on the evaluation of relaxation time (this will be discussed below) and involves the introduction of coordinates which move with the ion in its average drift through the solution.

The second reason for decrease in ionic mobility is the motion of the medium. This phenomenon, called the electrophoretic effect, is demonstrated by a well-known experiment: a porous container filled with tap water is partially immersed in a larger vessel also containing tap water, and when a potential difference is applied across the porous container, water is pumped through the wall. The accepted reason for this phenomenon is based on the idea that the water acquires a net charge when in contact with a surface: the wall either releases or adsorbs certain ions. When an electric field is applied, the net charges move and by means of friction they drag water along. In our electrolytic solution the atmosphere around each ion also consists of water with a net charge. When an external field is applied, the central ion finds itself moving in a medium which itself is drifting in the opposite direction; effectively, the central ion is always being pulled 'upstream'.

The magnitude of this effect may be calculated in first approximation [10] by neglecting the dissymmetry of the atmosphere, so that charge distribution is assumed to be spherically symmetrical and thus represented by the Boltzmann formula. If now we think of the atmosphere as consisting of a series of concentric spherical shells, we see that each shell is acted on by a net force equal to the product of its net charge and the external field. The net charge in each shell arises from the contained ions, and the resulting net force is transmitted to the liquid medium by means of viscous drag. If it is assumed that Stokes' law is valid, the liquid in each shell should move with a net velocity which is dependent on the net force, the viscosity of the liquid, and the radius of the shell. The velocity of the liquid in the immediate vicinity of the central ion is obtained by integrating all the velocity increments from the outermost shell to the central ion. According to this line of reasoning, the electrophoretic effect should vary with the charge and concentration of the ions, and in the calculations that have been made, this dependency does appear implicitly in the parameter κ.

In the steady state, then, it is not the simple equation $qE = \rho v$ which applies. Without going into the detailed derivation, we give the final expression for mobility which includes the correction for both atmospheric dissymmetry and the electrophoretic effect. In first approximation this is:

$$qE - \rho v \frac{e^2/\varepsilon\lambda}{6kT} = \rho v + qE\frac{a}{\lambda},$$

where a is the ionic radius, and all other symbols have their usual

References p. 251

meaning. In another form this gives a direct expression for velocity:

$$v = \frac{qE}{\rho}\left[1 - \frac{a}{\lambda} - \frac{e^2}{6\varepsilon kT}\frac{1}{\lambda}\right].$$

The parameter λ is a function of the square root of concentration, and so, according to this equation, ionic velocity and therefore conductivity vary linearly with the square root of concentration. Thus this equation represents a conductivity theory for dilute solutions which is in functional agreement with the experimental findings of Kohlrausch.

Relaxation Time

Of particular interest is the order of magnitude of the relaxation time, and this can be calculated using ideas of diffusion. The problem may be defined as follows: an ionic atmosphere has been built up around a central charge. This central charge is removed and the atmosphere decays through diffusion. How long does this take? Let n_i be the density of particles of type i. The particle current density J arises out of Brownian motion and electrical interactions between particles, and must obey the equation of continuity, $\partial n/\partial t = -\operatorname{div}\tilde{J}$. In the discussion of diffusion given in Chapter 3 it was shown how the particle current density which arises out of Brownian motion can be expressed in terms of the spatial variation of particle density; the equation is $\tilde{J} = -D \operatorname{grad} n$, where D is diffusion constant and can be written as kT/ρ. Current density resulting from electrical interactions we have already written as $\tilde{J} = nq\tilde{E}/\rho$. If field strength is written as the gradient of a potential Φ, this becomes: $\tilde{J} = -nq \operatorname{grad}\Phi/\rho$. With both of these effects taken into account, the expression for particle current density is:

$$\tilde{J} = \frac{-nq \operatorname{grad}\Phi}{\rho} - \frac{kT}{\rho}\operatorname{grad} n\,.$$

In a system containing two types of ions, there are at some distance from the central charge, n_1 positive and n_2 negative ions. The timewise change in concentrations n_1 and n_2 amounts to:

$$\frac{\partial n_1}{\partial t} = - \operatorname{div} \left[\frac{- q_1 \operatorname{grad} \Phi}{\rho_1} n_1 - \frac{kT}{\rho_1} \operatorname{grad} n_1 \right];$$

$$\frac{\partial n_2}{\partial t} = - \operatorname{div} \left[\frac{- q_2 \operatorname{grad} \Phi}{\rho_2} n_2 - \frac{kT}{\rho_2} \operatorname{grad} n_2 \right].$$

We know also that at every moment the potential Φ must meet the requirement of Gauss's law that:

$$\operatorname{div} \operatorname{grad} \Phi = \frac{-4\pi}{\varepsilon} \left[n_1 q_1 + n_2 q_2 \right].$$

These last three equations, taken with the proper boundary conditions, determine the solution of interest to us. Of course, when time variations vanish, these equations reduce to Boltzmann's law.

The differential equations just derived are non-linear, for they contain the product $n \operatorname{grad} \Phi$, and solution cannot be obtained as they stand. But we can at least get a proper order of magnitude if we make a reasonable first approximation: because the electrical potential energy is assumed much smaller than the thermal energy, the changes in concentration are small, and so we can replace the product $n_i \operatorname{grad} \Phi$ with $n_i^0 \operatorname{grad} \Phi$, where n_i^0 is the average concentration. If we restrict this discussion to the case in which the ions are of equal and opposite charge, e, the average concentration n^0 is the same for both types of ions. The additional assumption is made that the friction constants ρ_1 and ρ_2 are the same. With these assumptions the set of equations for which we want a solution becomes:

$$\frac{\rho}{kT} \frac{\partial n_1}{\partial t} = \operatorname{div} \left[\frac{n^0 e}{kT} \operatorname{grad} \Phi + \operatorname{grad} n_1 \right];$$

$$\frac{\rho}{kT} \frac{\partial n_2}{\partial t} = \operatorname{div} \left[\frac{- n^0 e}{kT} \operatorname{grad} \Phi + \operatorname{grad} n_2 \right];$$

$$\operatorname{div} \operatorname{grad} \Phi = - \frac{4\pi e}{\varepsilon} (n_1 - n_2).$$

References p. 251

We are interested primarily in deviations from uniform concentration, and so we write as a first approximation $n_1 = n^0 - \nu$ and $n_2 = n^0 + \nu$, where ν is a small variation. In terms of the new variable ν, the first two equations yield the same result:

$$\frac{\rho}{kT}\frac{\partial \nu}{\partial t} = \operatorname{div}\left[\frac{-n^0 e}{kT}\operatorname{grad}\Phi + \operatorname{grad}\nu\right],$$

and the third equation becomes:

$$\operatorname{div}\operatorname{grad}\Phi = \frac{8\pi e}{\varepsilon}\nu .$$

In the first of these new equations there is the term div grad Φ; the second equation is substituted here and the final result is a single differential equation in ν which combines all our information:

$$\frac{\rho}{kT}\frac{\partial \nu}{\partial t} = -\frac{8\pi n^0 e^2}{\varepsilon kT}\nu + \operatorname{div}\operatorname{grad}\nu .$$

The boundary conditions which apply to this problem are:
(a) At $t = 0$,

$$\nu = \text{const.}\frac{e^{-\varkappa r}}{r}\ * \ ;$$

(b) At $t = \infty$, the decay process is complete and $\nu = 0$.

The equation and boundary conditions can be put into a simpler form. We recognize the factor $8\pi n\, e^2/\varepsilon kT$ as κ^2. Also, we know that for our spherically symmetrical problem the operator div grad can be expressed as

$$\frac{1}{r^2}\frac{d}{dr}\left(r^2\frac{d}{dr}\right).$$

* By reasoning of the theory for strong electrolytes, we obtained the expression

$$\Phi = \text{Const.}\frac{e^{-\varkappa r}}{r},$$

and since ν is proportional to Φ,

$$\nu = \text{const.}\frac{e^{-\varkappa r}}{r}.$$

If in addition we express the distance r in terms of the dimensionless quantity s, defined by $\kappa r = r/\lambda \equiv s$, the differential equation becomes:

$$\frac{\rho}{\kappa^2 kT}\frac{\partial v}{\partial t} = -v + \frac{1}{s^2}\frac{\partial}{\partial s}\left(s^2 \frac{\partial v}{\partial s}\right).$$

By inspection, the quantity $\rho/\kappa^2 kT$ is seen to have the dimension of time, and must be identified with what we have termed relaxation time. So we define relaxation time as:

$$\theta = \rho/\kappa^2 kT.$$

The differential equation can be put into a dimensionless form by setting $t/\theta = \tau$. Then:

$$\frac{\partial v}{\partial \tau} = \frac{1}{s^2}\frac{\partial}{\partial s}\left(s^2 \frac{\partial v}{\partial s}\right) - v.$$

The corresponding boundary conditions are:
(a) At $\tau = 0$,

$$v = v_0 \frac{e^{-s}}{s},$$

where v_0 is some constant.
(b) At $\tau = \infty$, $v = 0$.
The solution is:

$$v = v_0 \frac{e^{-s}}{s} \frac{1}{\sqrt{\pi}} \int_{\sqrt{\tau} - s/(2\sqrt{\tau})}^{\infty} e^{-\xi^2}\, d\xi.$$

We shall not give here the mathematical steps for obtaining this solution, but the fact that it satisfies the boundary conditions is easily verified. When $\tau = 0$, the value of the integral is $\sqrt{\pi}$ and $v(0) = v_0 e^{-s}/s$ as required. Also, when $\tau = \infty$, the integral vanishes and $v(\infty) = 0$, also as required.

We know that in considering the speed of a decay process, a convenient measure is the time required for depletion of the decaying substance to some fraction of its initial amount. In terms of our problem, it is convenient to use the time required for the density increment to decrease to one half of its initial value, because this situation is represented by the above equation when the limits of the integral range from zero to infinity. When the lower limit is 0, $\sqrt{\tau} - s/(2\sqrt{\tau}) = 0$, or $\tau = s/2$. In terms of the original variables, $\kappa r/2 = t/\theta$, or t (50% decay) $= \kappa r\theta/2$; we see that fifty percent decay occurs at different times for different distances r.

The concept of relaxation time immediately suggested a certain experiment, of historical interest, which lent support to the theoretical idea[11]. If a finite time is required for the atmosphere to form or decay, it seemed reasonable that an ion oscillating with a period shorter than the relaxation time would be surrounded by an atmosphere with little dissymmetry. This condition could be achieved in a conductivity experiment by applying, instead of a direct voltage, an alternating voltage whose period is shorter than the relaxation time: the ions would merely oscillate slightly about their equilibrium positions. The 'effective' field would approach the applied field, and higher conductivity measurement would be expected. More specifically, conductivity should show a frequency dependence and should reach a maximum value at a frequency comparable in period to the relaxation time. The range of this required frequency is obtained from our expression for relaxation time, $\theta = \rho/\kappa^2 kT$. The friction constant ρ can be calculated using Stokes' law*; for an ion of radius 2Å in water of viscosity 10^{-2} cgs units, $\rho = 0.4 \times 10^{-8}$ cgs units. The thickness of ionic atmosphere in a one millimolar solution of uni-univalent electrolyte is 100 Å. Hence θ is of the order of 10^{-7} seconds (the time interval is so short because the forces involved are relatively quite large). Corresponding to this period is a frequency of 10^7 cycles per second, or a wavelength of 30 meters. When the indicated conductivity experiment was performed, frequencies in this range were used, and the anticipated dispersion of conductivity was found. Moreover, the frequency dependence of conductivity was found to be in complete accord with predictions of the theory.

Another conductivity experiment, not originally suggested by the idea of relaxation time but well explained by it, was performed by Max Wien.[12] In this work with ionic solutions, the dependence of conductivity on field strength was measured. It was found that as the applied field was increased from zero, conductivity remained virtually constant over a large range

* The validity of this law has been fairly well established for ions of medium size (larger than Li^+) through mobility experiments; at any rate, only an order of magnitude is needed here.

but showed big increases in the region of tens of kilovolts per centimeter. The explanation is that at high enough field strengths, the ions move with a velocity so great that they cannot build up atmospheres around them. The magnitude of field required to produce this effect can be estimated as follows. If we use the simple expression $\rho v = eE$, the velocity of a univalent ion whose friction constant is 0.4×10^{-8} cgs is calculated to be $4 \times 10^{-4}\ E$ (cm/sec), with E expressed in esu. The distance covered by the ion in the time required to build up a significant atmosphere is $\Delta x = v\theta$, and if θ is around 10^{-7} seconds, as in our example above, then $\Delta x = 4 \times 10^{-4} E \times 10^{-7}$ cm. When the field strength is sufficiently large so that this distance is equal to the thickness of the atmosphere, the ion will no longer be able to maintain a significant atmosphere and the conductivity should approach its value at infinite dilution. Such necessary field strength is calculated to be around 25 kilovolts per centimeter.

According to the idea of relaxation time, the value of conductivity at infinite dilution can be more closely approached through the use of high field strengths than by the alternating field method. The reason is that at high field strengths there is effectively no ionic atmosphere present, whereas at high frequencies, though the atmosphere has no dissymmetry, it is nevertheless present and gives rise to the electrophoretic effect.

APPENDIX TO CHAPTER 5

(*i*) *Calculation of Electrical Capacitance Including the Effect of Ionic Atmosphere*

The calculation given here is for a parallel plate condenser containing a dilute solution of electrolyte under conditions such that no discharge occurs at the plates.

Suppose there is a charge of magnitude σ per unit area of the electrode surface. Gauss's law can be applied to a small pillbox whose wall extends through the metal surface; one face lies in the solution, just outside the conductor, and the other lies within the metal where the field strength vanishes. If E is the electric field at the surface and ε is the dielectric constant, then $E = 4\pi\sigma/\varepsilon$. The equation relating electric field and potential is $E = -\,\mathrm{d}\Phi/\mathrm{d}x$. In terms of the dimensionless quantities κ and ξ, we have:

$$\frac{4\pi\sigma}{\varepsilon} = E = -\frac{\mathrm{d}\Phi}{\mathrm{d}x} = -\frac{kT}{e}\kappa\frac{\mathrm{d}\varphi}{\mathrm{d}\xi}.$$

Since:

$$\frac{\mathrm{d}\varphi}{\mathrm{d}\xi} = -\left(\mathrm{e}^{\varphi/2} - \mathrm{e}^{-\varphi/2}\right) = -\,2\sinh\frac{\varphi}{2},$$

then at the electrode surface where $\xi = 0$ and φ has the particular value v, we find:

$$\sigma = \frac{\varepsilon}{4\pi}\frac{kT}{e}\kappa\, 2\sinh\frac{v}{2}.$$

We shall consider the case in which v is much smaller than unity so that:

$$2\sinh(v/2) \simeq v$$

Since $v = eV/kT$:

$$\sigma = \frac{\varepsilon}{4\pi}\frac{kT}{e}\kappa\frac{eV}{kT} = \frac{\varepsilon}{4\pi}\frac{V}{\lambda},$$

in terms of familiar units. Accordingly, the capacitance per unit area is:

$$\frac{\sigma}{V} = \frac{\varepsilon}{4\pi\lambda}.$$

This is the expression for the capacitance of a parallel plate condenser if the distance λ is interpreted as the separation distance between the plates. Even in the case of a dilute solution, λ can be as small as 10^{-6} cm, which produces a capacitance much higher than that expected from just geometrical considerations.

At high potentials when v is not small compared to unity, one must use the expression 2 sinh (v/2), and charge density is seen to increase more rapidly than potential. If a large direct voltage is applied, and a small alternating voltage is superimposed in order to measure the capacitance, then the capacitance is:

$$C = \frac{\mathrm{d}\sigma}{\mathrm{d}V} = \frac{\varepsilon}{4\pi\lambda}\cosh\frac{v}{2} = C_0\cosh\frac{v}{2} = C_0\cosh\frac{eV}{2kT},$$

where C_0 is the capacitance per unit area at small potentials. For example, if the potential difference between the electrode and the solution were 100 mV, the effective capacitance as seen by a small alternating voltage superimposed on the large direct voltage is calculated to be C_0 cosh 2 or about 4 C_0; *i.e.*, it is four times larger than the low voltage capacitance.

(*ii*) *Correction to Van 't Hoff's Law Based on Dissociation Theory*

Consider the equilibrium M $\leftrightharpoons$ ν_1A + ν_2B. If n_0 represents the number of molecules before dissociation, α the degree of dissociation, and K the equilibrium constant, then:

$$\begin{aligned} (1-\alpha)n_0 &\equiv \text{number of molecules in solution;} \\ \nu_1\,\alpha\,n_0 &\equiv \text{number of A ions;} \\ \nu_2\,\alpha\,n_0 &\equiv \text{number of B ions.} \end{aligned}$$

According to Ostwald's law, at equilibrium:

$$K = \frac{(\nu_1\,\alpha\,n_0)^{\nu_1}\,(\nu_2\,\alpha\,n_0)^{\nu_2}}{(1-\alpha)n_0}.$$

References p. 251

If we define the sum $\nu_1 + \nu_2$ as ν, then Ostwald's law becomes:

$$\frac{\alpha^\nu}{1-\alpha} = \frac{K'}{{n_0}^{\nu-1}},$$

with K' some constant. At very low concentrations the dissociation is almost complete, and so we set $\alpha = 1 - \eta$, and treat η as small. To a good approximation, then, $\eta = {n_0}^{\nu-1}/K'$. The total number of particles is:

$$n = (1-\alpha)n_0 + \nu\,\alpha\,n_0 \left(1 - \frac{\nu-1}{\nu}\eta\right).$$

In these terms Van 't Hoff's law is:

$$\Pi = nkT = \nu n_0 kT \left(1 - \frac{\nu-1}{\nu}\eta\right).$$

Hence:

$$f = \frac{\nu-1}{\nu}\eta = \frac{\nu-1}{\nu K'}{n_0}^{\nu-1}.$$

A graph of f *vs.* concentration would appear as shown in Fig. 55. Curves a, b, and c are for the cases in which two, three, and four ions make up the molecule; f varies as c, c^2, and c^3 respectively.

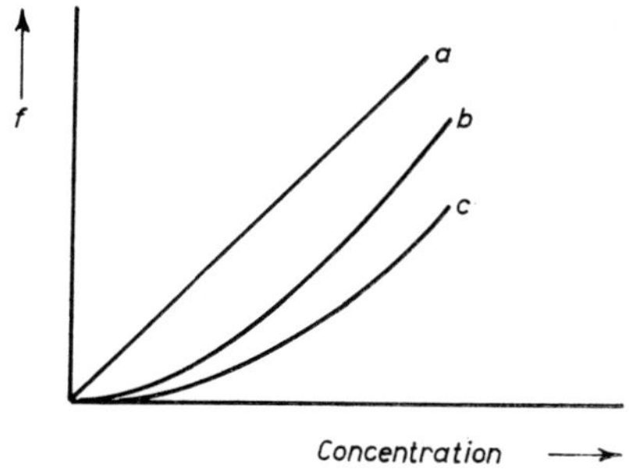

Fig. 55

(*iii*) *Interaction Energy Requirement for a Uniform Particle Distribution*

Uniform particle distribution is possible only if interaction energies are limited to those which arise from short-range forces. This conclusion can be reached by consideration of the equation for average potential energy per particle. For an unlimited volume of material:

$$\overline{\Phi} = \int_a^\infty \varphi(r) n 4\pi r^2 \,\mathrm{d}r \,,$$

where $\varphi(r)$ is the potential energy of one particle (assuming that φ is a function of distance only), n is density expressed as particles per cubic centimeter, r is distance from the central particle, and the limit a is the distance of closest approach. The question is under what circumstances n can be considered constant. If n were constant it could be removed from the integral, and since Φ must be finite, this is possible only if $\varphi(r) = A/r^{3+\delta}$; *i.e.*, if the potential energy of a particle falls off more rapidly than r^{-3}. Such energies arise from short-range forces, such as those occurring with uncharged particles. If long-range forces are involved, as for example in an ionic solution, it is not possible to assume a distribution characterized by a constant density. An extreme illustration is a gas composed of like ions: in such a gas the particles would distribute so that all the charge resides on a surface.

REFERENCES

1. ARRHENIUS, S., *Z. physik. Chem.*, 1, (1887) 631.
2. OSTWALD, W., *Z. physik. Chem.*, 2, (1888) 136, 276; 3, (1889) 170.
3. LEWIS, G. N., *J. Am. Chem. Soc.*, 34, (1912) 1631.
4. DEBYE, P. and HÜCKEL, E., *Physik. Z.*, 24, (1923) 185.
5. LEWIS, G. N. and RANDALL, M., *J. Am. Chem. Soc.*, 43, (1921) 1112.
6. ONSAGER, L., *Chem. Revs.*, 13, (1933) 73.
7. BJERRUM, N., *Z. Elektrochem.*, 24, (1907) 259.
8. KOHLRAUSCH, F., and HOLBORN, L., *Das Leitvermögen der Elektrolyte*, Teubner, Leipzig, 1916.
9. ONSAGER, L., *Physik. Z.*, 27, (1926) 388; 28, (1927) 277.
10. DEBYE, P. and HÜCKEL, E., *Physik. Z.*, 24, (1923) 305.
11. DEBYE, P. and FALKENHAGEN, H., *Physik. Z.*, 29, (1928) 121, 401.
12. WIEN, M., *Ann. Physik*, 83, (1927) 327.

CHAPTER 6

Fundamental Aspects of Coiling Molecules

Certain very large molecules, such as rubber or polystyrene, are made up of a great number of monomer units fastened together like links in a chain. In the study of solutions of such polymers, it is useful to know not only about the molecular weights of the molecules, but about how they are extended in solution. A measure of this extent is the average end-to-end distance, and we shall discuss theoretical and experimental ways of estimating it.

Each link in the polymer chain is free to rotate about its ends to some extent. In solution, Brownian motion causes the chains to assume more or less random configurations, and they are called 'random coils'. Of course, a number of factors affect the manner of coiling which occurs with a given polymer in a given solvent. The rotational freedom of the links is restricted by certain conditions: because of the chemical nature of the molecule, certain bond angles must be maintained, and because of the physical size of the chain there may be steric hindrances to free rotation. In addition, intermolecular forces and interactions may be significant. But in our discussion here we shall limit ourselves to a consideration of very dilute solutions, so that interactions may be reasonably neglected, and we shall assume for the most part that the links are completely free to rotate.

Let us consider, then, a rather fictitious molecule made up of links that are completely free to rotate about their ends; that is to say, if we consider one end of a given link to be fixed, the other end may lie anywhere on the surface of a sphere drawn about the fixed end. We suppose the chain is made up of N links, with N a large number; and we let the length and direction of each link be described by the vector $\tilde{r}_n$. All vectors are equal in magnitude, of course, but point in random directions. The definition of randomness means that the projection of the nth link on any arbitrary fixed axis, averaged over all the molecules, vanishes. Similarly, the projection of the vector $\tilde{V}$, which proceeds from end A of a molecule to end B, vanishes when averaged over all the molecules. This merely means that such a molecule has no preferred orientation in space.

The average of the square of the vector $\tilde{V}$ does exist, however, and is easily calculated. We know that $\tilde{V} = \sum_n \tilde{v}_n$; the square of this can be written $V^2 = \sum_m \sum_n \tilde{v}_m \cdot \tilde{v}_n$. The average value of V^2 is obtained by summing the average of each term in the summand:

$$\overline{V}^2 = \sum_m \sum_n \overline{\tilde{v}_m \cdot \tilde{v}_n} \,.$$

Let us first examine the expression $\overline{\tilde{v}_m \cdot \tilde{v}_n}$. This represents the scalar product of any specific pair of vectors within a coil, averaged over all the random coils. Each term is of the form $a^2 \cos \vartheta$, where a is the length of each vector and ϑ is the angle between them. If n and m are different, we can summarize the average $\overline{\tilde{v}_m \cdot \tilde{v}_n}$ simply as $a^2 \overline{\cos} \vartheta$. But since the links are absolutely independent, there is no preferred orientation and the average of the cosine vanishes. Hence $\overline{\tilde{v}_m \cdot \tilde{v}_n}$ vanishes for $m \neq n$. However, when m and n are the same (*i.e.*, when they refer to the same link within each coil), the angle ϑ is always zero and $\cos \vartheta$ is always unity. Thus $\overline{\tilde{v}_m \cdot \tilde{v}_n} = a^2$ when $m = n$. In sum we have:

$$\overline{V}^2 = \sum_n v_n{}^2 = Na^2 \,.$$

Average end-to-end distance, R, is defined as the square root of $\overline{V^2}$. Finally, then *:

$$R = \sqrt{N}\, a \,.$$

As would be expected, end-to-end distance is seen to increase much more slowly than the number of links; the important feature of this result is that the functional relation between R and N is shown to be a square-root dependency. When molecules are large, the effect of such a dependency is very great. For example, if we had a random coil made up of 10^4 links, each of length 1 mm, the chain when fully extended would measure

* This problem is mathematically identical to the so-called 'random flight' problem. When applied to a plane rather than to three-dimensional space, this is the familiar 'random walk' problem.

References p. 271

10 meters, yet under conditions of complete randomness its average end-to-end distance would be only 10 cm.

The calculation of end-to-end distance can be improved by basing it on a more realistic model in which it is required that certain bond angles be maintained. The stipulation is that one end of a given link must be located on a particular conical surface described about the other end, the axis of the cone being an extension of the adjacent link (see Fig. 56).

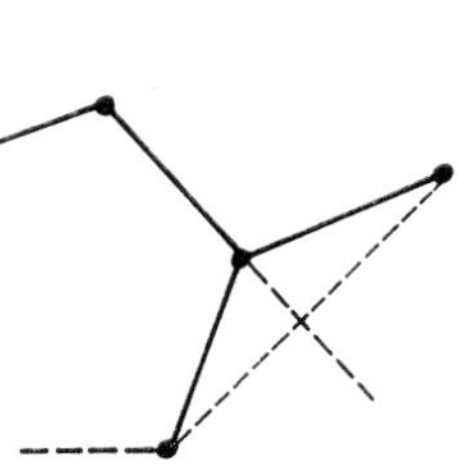

Fig. 56

Average end-to-end distance calculated on this basis is of the same form as that calculated with the assumption of free rotation: the difference is in degree only. If the chain is constructed of linked carbon atoms with their particular required bond angles, the calculation[1] shows $R = \sqrt{2N}\, a$.

We could further improve our model by imposing a more important restriction: because each monomer unit occupies a finite volume, the space available for any one monomer is limited by the space already occupied by all the others. However, the use of this restriction would require a solution of the so-called 'volume exclusion' problem* which has received much attention but has not been satisfactorily solved. Long-range effects are involved in such a volume-exclusion problem, and not even the form of the answer is known. But it seems probable that if we were to consider our random-coil calculation in terms of a volume-exclusion restriction, the functional dependence of end-to-end distance on number of links would be altered. (This is in contrast to the consequences of the bond-angle restriction, which exerts a very short-range effect.) Attempts

* The corresponding two-dimensional problem illustrates the volume-exclusion problem. The model is a planar grid, and beginning with a starting point located in any square, we take a series of N steps. Each step must be to an adjacent square, but stepping into a square previously occupied is forbidden. The problem is to find the average end-to-end distance.

have been made to solve volume exclusion problems numerically with the aid of calculating machines, but as yet calculators are not able to handle numbers as big as the number of links in a large random coil polymer.

Elongation of Random Coils

Let us return, then, to the simpler case in which volume exclusion is neglected, and investigate the elasticity of a random coil material. The elongation of such a material is an interesting phenomenon, because although it increases with the applied force, as would be expected, it shows the surprising property of decreasing as temperature rises. The reason for this behavior is that there is no potential energy of bond bending or stretching involved. Instead, the energy involved is kinetic, arising from the constant movement of the links under the influence of Brownian motion. Trying to stretch such a chain is like pulling on a child's jumping rope which is in motion: the force that must be counteracted is due to acceleration rather than to potential energy. In the case of our random coil, a temperature rise would increase the kinetic energy of the links, and so tend to shorten the chain.

In order to develop a mathematical expression for elongation, let us first decompose the oppositely directed applied forces acting on the ends of the coil into pairs of vector forces grouped to give the effect of torques acting on each link (see Fig. 57). Then, with reference to an axis drawn

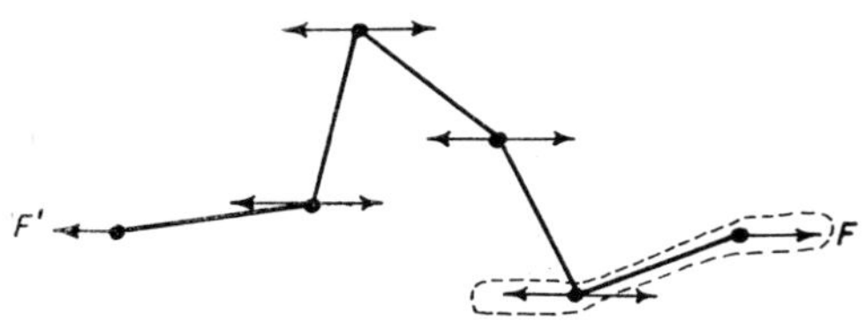

Fig. 57

along the direction of the forces F and F', each bond is under the influence of a torque of $-Fa \sin \vartheta$, where a is the length of each link and ϑ is the inclination of the link from the axis. Now we have pointed out that there is no potential energy involved in the elongation of the chain. However, the force which arises from the kinetic energy of the links may be treated as a force arising from a potential energy. The reasoning is as follows:

References p. 271

when the energy involved in an equilibrium is potential, the resultant torque is represented as the derivative $-\partial V/\partial \vartheta$; in the case of our coiling chain, the torque on a link equals $-Fa \sin \vartheta$, and so we may speak of a potential energy of the link as $V = -Fa \cos \vartheta$.

At any instant, the number of links whose directions lie within some solid angle $d\Omega$ is, by Boltzmann's law, proportional to [exp (— potential energy/kinetic energy) $d\Omega$], which in our case is $e^{Fa \cos \theta/kT} d\Omega$. If we consider the sphere described by the solid angle 4π, we see that the total number of links contained in the sphere is proportional to

$$\int_0^{4\pi} e^{Fa \cos \theta/kT} d\Omega .$$

Now we are interested in the projection of the links on the axis. For any one link, this is $a \cos \vartheta$, and the sum of projections for all links is proportional to $\int e^{Fa \cos \theta/kT} a \cos \vartheta \, d\Omega$. The average projection of a link equals the sum of projections of all the links contained in the sphere, divided by the total number of links. Hence the average projection, which may also be expressed as $(a \overline{\cos \vartheta})$, is, for one link:

$$a \overline{\cos \vartheta} = \frac{\int e^{Fa \cos \theta/kT} a \cos \vartheta \, d\Omega}{\int e^{Fa \cos \theta/kT} d\Omega} \equiv z .$$

An integral of similar form was evaluated in Chapter 1 in connection with a discussion of orientation of dipoles in an electric field; by similar procedures, z may evaluated to obtain the result:

$$z = a \left\{ \coth \frac{aF}{kT} - \frac{1}{aF/kT} \right\} = a \mathscr{L} \left(\frac{aF}{kT} \right),$$

where $\mathscr{L}$ is the Langevin function. Since the links are assumed independent, the elongation L is Nz, and so finally:

$$L = Na \mathscr{L} \left(\frac{aF}{kT} \right).$$

We can inspect this equation to see how reasonable it is in terms of observable behavior. If the forces involved are sufficiently small, a simple approximation can be made for the Langevin function: $\mathscr{L}(x) = x/3$.

Accordingly,

$$L = \frac{Na^2}{3kT} F ,$$

and we see that, as in Hooke's law, elongation is proportional to the force. The interesting feature, however, is the appearance of the term $1/T$, which indicates that with the same applied force, length decreases as temperature rises; as pointed out above, this is an observed property of such materials. The reasonableness of the equation for the case of large applied forces can also be assessed by inspection. At an infinite applied force, $\mathscr{L}$ is close to unity, and L approaches the value Na; this expresses the fact that a chain can extend no further than its total length.

Effect of Fluctuations on Average End-to-End Distance

We have already derived an expression for average end-to-end distance of a random coil molecule. However, with any chain composed of a finite number of links, we must expect fluctuations of end-to-end distance about this average value, and the shorter the chain length, the more significant we would expect these fluctuations to be. Suppose we have a chain with end A fixed, and we wish to consider all the possible positions of end B. We want to know not only its most probable or average position, but also the probability of finding end B at any position in space. This is equivalent to saying that we want to find the probability for all possible fluctuations of end-to-end distance about the average value. The mathematical treatment of this problem may be carried out in several ways, and we shall discuss in some detail a direct and an indirect method.

The indirect method features a technique often used in the field of statistics, the use of the Dirichlet discontinuous factor. To use this method, we first set up a probability equation. The probability that a given independent vector $\tilde{v}$ will point in a direction encompassed by the element of solid angle $d\Omega$ is $d\Omega/4\pi$. For a number of independent vectors, $\tilde{v}_1, \tilde{v}_2, \ldots \tilde{v}_N$, to point, respectively, in the directions designated by $d\Omega_1, d\Omega_2, \ldots d\Omega_N$, the probability is $d\Omega_1/4\pi \cdot d\Omega_2/4\pi \ldots d\Omega_N/4\pi$. We designate that end A is fixed and is located at the origin of Cartesian coordinates, and we want to calculate the probability that end B will lie in the element of volume whose projection on the z axis is in the range

ζ to $\zeta + \mathrm{d}\zeta$ (see Fig. 58). This means that the sum of individual link projections of the z axis should be between the values ζ and $\zeta + \mathrm{d}\zeta$; if ϑ_n represents the inclination of the nth link to the z axis, the total projection of the chain length is $z = a \sum_n \cos \vartheta_n$. We shall call the probability $P\,\mathrm{d}\zeta$, where P is the probability function. Thus:

$$P\,\mathrm{d}\zeta = \int\int \cdots \int_{a\Sigma_n \cos\vartheta_n = \zeta}^{a\Sigma_n \cos\vartheta_n = \zeta + \mathrm{d}\zeta} \frac{\mathrm{d}\Omega_1}{4\pi}\frac{\mathrm{d}\Omega_2}{4\pi} \cdots \frac{\mathrm{d}\Omega_N}{4\pi}\,.$$

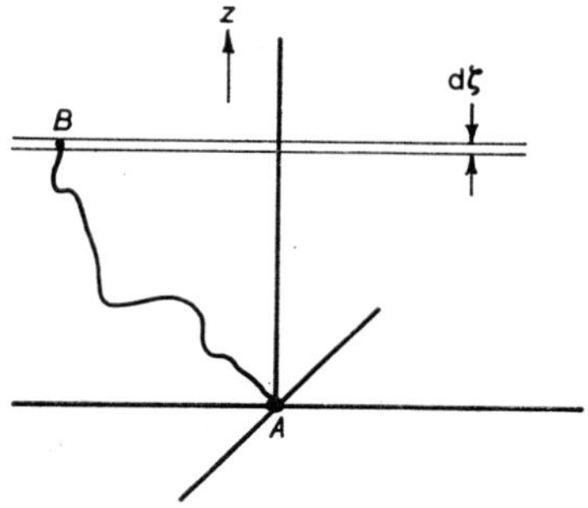

Fig. 58

This must be transformed to appropriate variables. Since the total chain length projection is described in terms of the polar angle ϑ_n, the element of solid angle may also be expressed in these terms as follows: $\mathrm{d}\Omega_n/4\pi = 1/2 \sin\vartheta_n\,\mathrm{d}\vartheta_n$. Then if we define $x_n \equiv \cos\vartheta_n$, $\mathrm{d}x_n$ equals $-\sin\vartheta_n\,\mathrm{d}\vartheta_n$, and the absolute magnitude of this expression differs from $\mathrm{d}\vartheta_n/4\pi$ only by the factor 1/2. The probability integral is then:

$$P\,\mathrm{d}\zeta = \frac{1}{2^N}\int\int \cdots \int_{a\Sigma_n x_n = \zeta}^{a\Sigma_n x_n = \zeta + \mathrm{d}\zeta} \mathrm{d}x_1\,\mathrm{d}x_2 \ldots \mathrm{d}x_N\,.$$

Interpreted geometrically, this equation means that we wish to find the volume, in N-dimensional space, which is included between two planes, one at $a \sum_n x_n = \zeta$ and the other at $a \sum_n x_n = \zeta + \mathrm{d}\zeta$.

We can now make use of the Dirichlet discontinuous factor, which may be represented graphically as a rectangle of unit height (see Fig. 59). The proper inclusion of such a factor in the integrand of our probability equation will allow us to integrate over the entire range of variation of our x's, because the discontinuous factor will automatically confine the limits to the required range.

First it is necessary to put this discontinuous factor, D, into convenient form. Since it is a piecewise smooth function in the range $-\infty < u < \infty$, it can be represented, according to the theory of Fourier integrals, as:

$$\mathrm{D}(u) = \int_{-\infty}^{+\infty} \mathrm{e}^{\mathrm{i}2\pi\nu u}\, \varphi(\nu)\, \mathrm{d}\nu ,$$

where

$$\varphi(\nu) = \int_{-\infty}^{+\infty} \mathrm{e}^{-\mathrm{i}2\pi\nu u}\, \mathrm{D}(u)\, \mathrm{d}u .$$

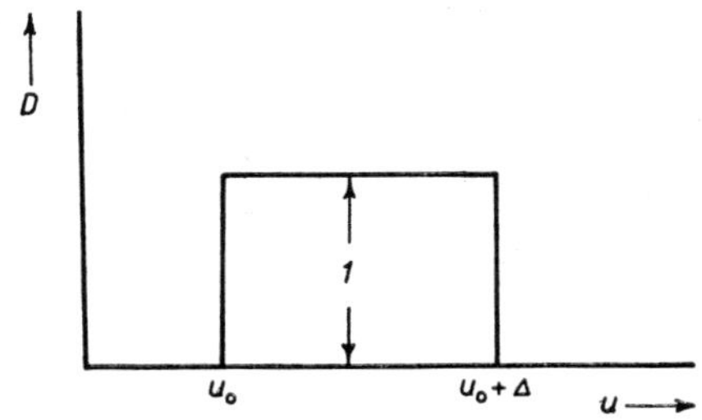

Fig. 59

We know that $\mathrm{D}(u) = 1$ when $u_0 \leqslant u \leqslant u_0 + \Delta$, and so we may write:

$$\varphi(\nu) = \int_{u_0}^{u_0+\Delta} \mathrm{e}^{-\mathrm{i}2\pi\nu u}\, 1\, \mathrm{d}u = \mathrm{e}^{-\mathrm{i}2\pi\nu u_0} \left\{ \frac{1 - \mathrm{e}^{-\mathrm{i}2\pi\nu\Delta}}{\mathrm{i}2\pi\nu\Delta} \right\} \Delta .$$

Upon substitution:

$$\mathrm{D}(u) = \int_{-\infty}^{+\infty} \mathrm{e}^{\mathrm{i}2\pi\nu\,(u - u_0)}\, \frac{1 - \mathrm{e}^{-\mathrm{i}2\pi\nu\Delta}}{\mathrm{i}2\pi\nu\Delta}\, \Delta\, \mathrm{d}\nu .$$

Now we must adjust the values of u_0 and Δ to our problem by identifying u with $a\sum_n x_n$, u_0 with ζ, and Δ with $\mathrm{d}\zeta$. Our probability equation can now be written:

$$P\,\mathrm{d}\zeta = \frac{1}{2^N}\int\limits_{x=-1}^{+1}\cdots\int\limits_{-1}^{+1}\int\limits_{\nu=-\infty}^{+\infty} \mathrm{e}^{\mathrm{i}2\pi\nu\,(a\Sigma_n x_n - \zeta)}\;\frac{1-\mathrm{e}^{-\mathrm{i}2\pi\nu\Delta}}{\mathrm{i}2\pi\nu\Delta}\;\Delta\,\mathrm{d}x_1\ldots\mathrm{d}x_N\,\mathrm{d}\nu\,.$$

The integration is carried out first over the x's. Each x term integrates as follows:

$$\frac{1}{2}\int\limits_{-1}^{+1}\mathrm{e}^{\mathrm{i}2\pi\nu a x_n}\,\mathrm{d}x_n = \frac{\sin 2\pi\nu a}{2\pi\nu a}\,.$$

With all the N terms involving x integrated, the probability equation becomes:

$$P\,\mathrm{d}\zeta = \int\limits_{-\infty}^{+\infty}\left(\frac{\sin 2\pi\nu a}{2\pi\nu a}\right)^N \frac{1-\mathrm{e}^{-\mathrm{i}2\pi\nu\Delta}}{\mathrm{i}2\pi\nu\Delta}\;\Delta\,\mathrm{e}^{-\mathrm{i}2\pi\nu\zeta}\,\mathrm{d}\nu\,.$$

In the limiting case in which values of Δ or $\mathrm{d}\zeta$ are small, the second term is unity. The remaining Δ cancels $\mathrm{d}\zeta$ on the left-hand side, and we obtain the expression for P alone:

$$P = \int\limits_{-\infty}^{+\infty}\left(\frac{\sin 2\pi\nu a}{2\pi\nu a}\right)^N \mathrm{e}^{-\mathrm{i}2\pi\nu\zeta}\,\mathrm{d}\nu\,.$$

This is exact for any N if the effect of excluded volume is neglected. Although as it stands the integral is difficult to evaluate, it is possible to use a great simplification if N is large. We can expand the first term in such a way as to get an approximation in exponential form, which can then be combined with the exponential term already present in the equation. The function $\sin x/x$ is an oscillating function which begins at unity at the origin and decreases in amplitude like $1/x$. The value of the Nth power of this function is also unity at the origin but it decays toward zero much more rapidly, for its amplitude decreases like $(1/x)^N$. Bearing this in mind, we use what is essentially a method of steepest descent to evaluate $(\sin x/x)^N$. First we expand the logarithm of this function in a

Taylor series: $N \ln(\sin x/x) = N \ln(1 - x^2/6 + \ldots)$. For small values of x this reduces to $-(x^2/6)\,N$. Taking antilogarithms we have: $(\sin x/x)^N \simeq e^{-Nx^2/6}$. Of course, this is valid only in a region around the origin, but since the function falls off towards zero so rapidly, this approximation is accurate enough. Thus the equation for P becomes:

$$P = \int_{-\infty}^{+\infty} e^{-N\frac{4\pi^2\nu^2a^2}{6}}\, e^{-i2\pi\nu\zeta}\, d\nu\,;$$

or, with $\mu = 2\pi\nu$:

$$P = \frac{1}{2\pi}\int_{-\infty}^{+\infty} e^{-\frac{Na^2}{6}\mu^2}\, e^{-i\mu\zeta}\, d\mu\,.$$

The exponent is treated by the method of completion of squares to give:

$$\frac{Na^2}{6}\mu^2 + i\mu\zeta = \frac{Na^2}{6}\left[\mu + \frac{i3\zeta}{Na^2}\right]^2 + \frac{3\zeta^2}{2Na^2}\,.$$

When the exponent is used in this form in the integral, our equation becomes:

$$P = e^{-3/2\,(\zeta^2/Na^2)}\,\frac{1}{2\pi}\int_{-\infty}^{+\infty} e^{-\frac{Na^2}{6}[\mu+\mathrm{const.}]^2}\, d\mu\,.$$

The integral is of the form

$$\int_{-\infty}^{+\infty} e^{-\alpha x^2}\, dx\,,$$

which is evaluated as $(\pi/\alpha)^{1/2}$. Finally:

$$P = \left(\frac{3}{2\pi Na^2}\right)^{1/2} e^{-3/2\,(\zeta^2/Na^2)}\,,$$

where P is a normalized quantity. It should be noted that this expression is valid only when N is large; also, there is included no correction for any rotational restriction, and, as in all our calculations, the effect of excluded volume is neglected.

References p. 271

The direct method of deriving a probability expression proceeds along entirely different lines. We again suppose that end A is fixed at the origin; then if ds is some volume element in space, the probability that end B will be found within ds is f ds. The probability function f depends on the three space variables as well as on N, the number of links between A and B.

We begin with the supposition that this function is already known for N links, and on this basis determine the effect on f of adding one more link. In other words, we derive an expression for f_{N+1} in terms of f_N. If the end of link $(N + 1)$ is to lie somewhere within the small spherical element of size ds, then the Nth end must lie somewhere within a spherical shell described about the central volume element (see Fig. 60). The

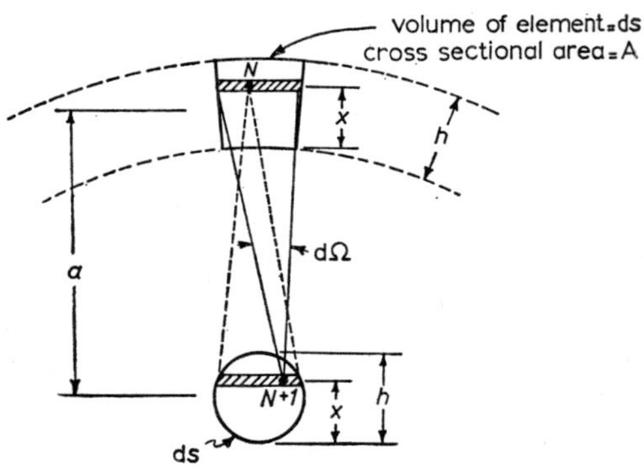

Fig. 60

thickness of the shell, h, equals the diameter of the central volume element, and the average radius of the shell equals the link length, a. The probability that the end of link $(N + 1)$ lies within the central volume element is the composite probability that the end of link N lies within the spherical shell and, simultaneously, that link $(N + 1)$ is oriented in the direction of the central volume element. Let us consider a cylindrical element of volume in the spherical shell, of height h, cross-sectional area A, and volume, hA, equal to that of the central element ds. First of all, we can say that the probability that the end of link N lies within this shell element is $f_N{}^*$ ds, where the asterisk refers to the coordinates of the shell. If end N lies within the shell element at a distance x from the base, then if end $(N + 1)$ is within the central volume element ds, it must lie within a sheet located at the same distance, x, above the lowest point. The area of this sheet increases from zero when x is zero, to a maximum when $x = h/2$, and then decreases to zero when $x = h$. If we speak of the occupancy of the shell element as a whole, then we may speak of an average area in which end $(N + 1)$ must lie: this average area is equal, by definition, to A, the cross-sectional area of the shell element. If we translate this condition involving area to one involving solid angle, we may say that if end N lies within the shell volume element, then in order for end $(N + 1)$ to lie within ds, it must lie within the solid angle A/a^2. But this solid angle is precisely equal to the solid angle subtended by the shell element from end $(N + 1)$ in the central volume element. If we call this solid angle dΩ, then with the assumption of free rotation, the total probability for end $(N + 1)$ to be located in the central volume can be written:

$$f_{N+1}\,\mathrm{d}s = \int_0^{4\pi} f_N{}^*\,\mathrm{d}s\,\frac{\mathrm{d}\Omega}{4\pi} = \mathrm{d}s \int_0^{4\pi} f_N{}^*\,\frac{\mathrm{d}\Omega}{4\pi}\,.$$

The last integral represents the average value of $f_N{}^*$ over the sphere, and so:

$$f_{N+1} = \overline{f_N{}^*}\,.$$

These probability functions depend, as we pointed out above, on the three space variables as well as on the number of links. Now we must designate some origin to which we can refer all probabilities, and for this origin we choose the location of the volume element containing end $(N + 1)$. Points on the spherical shell are accordingly located by the coordinates ξ, η, and ζ. Then the average value of $f_N{}^*$ is, by Taylor expansion:

$$\overline{f_N{}^*} = f_N + \left(\bar{\xi}\,\frac{\partial f_N}{\partial x} + \bar{\eta}\,\frac{\partial f_N}{\partial y} + \bar{\zeta}\,\frac{\partial f_N}{\partial z}\right) + \frac{1}{2}\left(\overline{\xi^2}\,\frac{\partial^2 f_N}{\partial x^2} + \ldots\right) + \ldots\,,$$

where the right-hand side is evaluated at the origin. No further terms need be evaluated, since for a molecule with a large number of links the function f varies so little over the distances being considered. Because of our assumption that volume exclusion can be neglected, we may say that the averages $\bar{\xi}$, $\bar{\eta}$, $\bar{\zeta}$, $\overline{\xi\eta}$, $\overline{\xi\zeta}$, and $\overline{\eta\zeta}$ all vanish, and furthermore that the averages $\overline{\xi^2}$, $\overline{\eta^2}$, and $\overline{\zeta^2}$ must all equal $a^2/3$. The Taylor expansion thus

References p. 271

becomes:

$$f_{N+1} = f_N + \frac{a^2}{6} \text{ div grad } f_N \,.$$

This is a difficult equation to solve, since it is a difference equation in the number of links, N, and a differential equation in x, y, and z. However, it is apparent that when N is large, f_{N+1} and f_N are very nearly the same, so that we assume:

$$f_{N+1} - f_N = \frac{\partial f}{\partial N} = \frac{a^2}{6} \text{ div grad } f \,.$$

This equation is identical in mathematical form to the diffusion equation, and so the solution to the diffusion equation is immediately applicable to our problem. In terms of our variables the solution is:

$$f = \left(\frac{3}{2\pi N a^2}\right)^{3/2} \mathrm{e}^{-3/2\,(r^2/Na^2)} \,,$$

where r is the distance between the chain ends.

For the purpose of comparison with the expression already obtained by the indirect method, the factors of f may be regrouped:

$$f = \left(\frac{3}{2\pi N a^2}\right)^{1/2} \mathrm{e}^{-3/2\,(\xi^2/Na^2)} \cdot \left(\frac{3}{2\pi N a^2}\right)^{1/2} \mathrm{e}^{-3/2\,(\eta^2/Na^2)} \cdot \left(\frac{3}{2\pi N a^2}\right)^{1/2} \mathrm{e}^{-3/2\,(\zeta^2/Na^2)} \,.$$

Here, each term represents the chances of finding a particular projection of end-to-end distance on one of the three axes x, y, or z. The last term, involving ζ, is identical to that derived by the indirect method. We may, though, call attention to a potential advantage of the direct method: it may prove to be useful in attacking the volume exclusion problem, since it might be generalized. The difficulty lies in going to a differential equation from a difference equation.

The function f, we recall, is the normalized probability function which, when multiplied by a volume element, gives the probability of finding a chain length in the space. Average end-to-end distance for the simple model we have assumed can be obtained from the general probability expression, for in accordance with definitions already established we may write:

$$R^2 = \overline{r^2} = \int_0^\infty \left(\frac{3}{2\pi Na^2}\right)^{3/2} e^{-3/2\,(r^2/Na^2)}\, r^2\, 4\pi r^2\, dr\,\dagger\,.$$

This yields the expression $R = \sqrt{N}a$, which is identical to the expression derived earlier in a simple manner using the definition of complete randomness. It seems reasonable, and in fact it has been shown, that if one wishes to apply the probability expression to a problem which involves short-range forces (as, for example, in the maintenance of certain bond angles), one can simply replace Na^2, wherever it appears, with the corrected value for the square of average end-to-end distance*.

As conclusion to our discussion of probability function, let us now briefly consider what information it provides and for what purposes it is meaningful. We must remember that any mathematical expression for probability is based on an assumed model, and therefore that its application to the prediction or interpretation of physical measurements is limited by the validity of the model. Assuming that the model sufficiently resembles the structure of the molecule, the probability function provides a characterization of the density distribution. This information can be used to calculate certain physical constants or properties of the dissolved particles

† The integral can be put into the form

$$\int_0^\infty e^{-\alpha x^2}\, x^4\, dx\,.$$

We already know that

$$\int_0^\infty e^{-\alpha x^2}\, dx = (\sqrt{\pi}/2)\,\alpha^{-1/2}\,;$$

we differentiate both sides of this equation twice with respect to α to obtain

$$\int_0^\infty e^{-\alpha x^2}\, x^4\, dx = \left[\frac{3\sqrt{\pi}}{8}\right] \alpha^{-5/2}\,.$$

Simple substitution completes the evaluation.

* For a discussion of this point, see PAUL J. FLORY's book, *Principles of Polymer Chemistry* [2].

or the solutions. An example is the calculation, given below, of average end-to-end distance from light scattering measurements.

Estimation of End-to-End Distance by Means of Light Scattering Measurements

How can we measure experimentally the extension of a random coil in solution? It has already been pointed out, in Chapter 4, that the amount of polarized light scattered by particles small in comparison to the wavelength is independent of the angle of observation (in an equatorial plane). If, however, the particles are not small compared to the wavelength, destructive interferences within the particle give rise to more forward than backward scattering. Thus if we illuminate a polymer solution with light whose wavelength is comparable to the average end-to-end distance of the molecules, the scattering which we observe should exhibit a large amount of angular dissymmetry, and this should be a measure of particle length. Of course, the measurement must be interpreted in terms of a molecular model, for which we shall take a simple random coil.

Consider, for example, polystyrene of average molecular weight 10^6, dissolved in a solvent with which its interactions are such as to make the random coil ideas appropriate. The monomer weight is about 100, and so the average number of monomer units per coil is 10^4, and the average number of links is 2×10^4. Each link is 1.54 Å long; hence the average end-to-end distance calculated according to the expression $R = \sqrt{2N}a$ is 308 Å. Unfortunately, light of such wavelength cannot be used in this study because it is too strongly absorbed, but we should expect significant dissymmetry even at wavelengths near the limit of the visible spectrum, around 3000 Å.

The equation for R was derived from a simple random coil model, with no corrections made for excluded volume effects which give rise to greater end-to-end distance. In addition, the extending effect of interactions between solvent and solute were neglected. We should, then, expect that end-to-end distance calculated from light scattering measurements would be higher than that calculated on the basis of the simple model. Indeed, in benzene solutions of polystyrene, where there is much interaction, the measured dissymmetry corresponds to an average end-to-end distance of 1100 Å, about four times the theoretical value.

Let us see how to calculate average end-to-end distance of random coil particles in solution from measurements of light scattering. First we shall need to review the light scattering theory, and revise it by using the random coil probability function to more accurately characterize the structure of the particle.

As shown in Chapter 4, the amplitude of light scattered by a particle is given by:

$$A \propto \sum_n \mathrm{e}^{\mathrm{i}k(\tilde{s}\cdot\tilde{r}_n)},$$

where r_n locates the nth scattering center within the particle, k represents the quantity $2\pi/\lambda$, and $\tilde{s}$ is the vector difference between the unit vectors $\tilde{S}_0$ and $\tilde{S}$, which represent respectively the direction of the primary beam and the direction of the path from the particle to the point of observation. For any one spatial configuration of the particle, the intensity of scattered light is given by:

$$I = (\mathrm{Re}\,A)^2 = \frac{1}{2}AA^* \propto \sum_n \sum_m \mathrm{e}^{\mathrm{i}k\tilde{s}\cdot\tilde{r}_{mn}},$$

where $\tilde{r}_{mn}$ stands for $\tilde{r}_m - \tilde{r}_n$. With the passage of time, the coil assumes many configurations, and so the actual intensity observed is the average value of I:

$$I_{\text{obs.}} = \sum_m \sum_n \langle \mathrm{e}^{\mathrm{i}k\tilde{s}\cdot\tilde{r}_{mn}} \rangle.$$

According to this, we sum all the possible interactions between particles, considering the particles two at a time.

The averaging process may be performed in two steps. First, for a given pair mn, let us consider the vector $\tilde{r}_{mn}$ whenever it is of some particular length and average over all its possible directions. Since the vector $\tilde{s}$ is constant, and since there is no preferred orientation of $\tilde{r}_{mn}$, this average value is:

$$\frac{1}{4\pi}\int_0^\pi \mathrm{e}^{\mathrm{i}ksr\cos\theta}\, 2\pi \sin\vartheta\, \mathrm{d}\vartheta = \frac{\sin(ksr)}{ksr},$$

in which only scalar quantities are involved. Our intensity equation thus becomes:

References p. 271

$$I_{\text{obs.}} \propto \sum_m \sum_n \left\langle \frac{\sin (ksr_{mn})}{ksr_{mn}} \right\rangle .$$

The second step in evaluating the averaged quantity involves the consideration of variations in vector length. Therefore, we multiply the term $\sin (ksr_{mn})/ksr_{mn}$ by the probability that the separation distance r_{mn} lies between r and $r + \mathrm{d}r$. Upon integration from $r = 0$ to $r = \infty$, we obtain the average value that applies to these two particular particles which are separated by a number of links ν. Thus the average pertaining to a particular value of ν is:

$$\int_0^\infty \frac{\sin (ksr)}{ksr} \left[\frac{3}{2\pi \nu a^2}\right]^{3/2} \mathrm{e}^{-3/2\,(r^2/\nu a^2)}\, 4\pi r^2\, \mathrm{d}r .$$

We can obtain a form suitable for integration by identifying $\sin (ksr)$ with $Im\,(\mathrm{e}^{ikrs})$; the resulting expression involves exponentials only. The square is completed and integration proceeds smoothly. Upon extracting the imaginary part of the answer, we obtain the average, $\mathrm{e}^{-k^2s^2\nu a^2/6}$. It should be noted that the value of this average quantity (and therefore of intensity of light scattering) decreases with increasing number of links, which is a condition to be expected since the greater the number of links between two particles, the more nearly independent will be their motions in space. For convenience, the exponential is abbreviated as $\mathrm{e}^{-\alpha\nu}$, where $\alpha = (1/6)(4\pi/\lambda^2)(s^2a^2)$. We note that since the magnitude of s can never be greater than two, and since the ratio a^2/λ^2 is a very small number at optical frequencies, the value of α must be very small. Using the abbreviated form for the averaged quantity in our expression for observed intensity, we obtain:

$$I_{\text{obs.}} \propto \sum_m \sum_n \mathrm{e}^{-\alpha\nu} .$$

Next the double summation must be performed. To do this, we begin with some initial particle, and express the interaction between it and all the other particles as the sum of the above exponential terms over all ν, which is the number of links between this and any other particle. Then another initial particle is chosen and the process repeated. This continues until all N particles have had a turn at being the initial particle, and then the N sums are added together to give a grand sum which represents the desired double summation. Following this plan, we begin by writing for the result of interaction between the first particle and the rest:

$$1 + e^{-\alpha} + e^{-2\alpha} + \ldots + e^{-(N-1)\alpha} .$$

The result of interactions between the second particle and all the rest is:

$$e^{-\alpha} + 1 + e^{-\alpha} + e^{-2\alpha} + \ldots + e^{-(N-2)\alpha} .$$

Similarly, the evaluation for observed intensity resulting from interactions between the third particle and all the rest is:

$$e^{-2\alpha} + e^{-\alpha} + 1 + e^{-\alpha} + e^{-2\alpha} + \ldots + e^{-(N-3)\alpha} .$$

The entire sum includes N such terms.

We note first of all that if $\alpha = 0$ (this corresponds to the case of a point particle), $I \propto N^2$. This is as expected when there are no interferences: amplitude is proportional to number of particles, and intensity is the square of amplitude. For the more general case of $\alpha \neq 0$, we note that the diagonal terms sum to N, so that:

$$I \propto N + 2(N-1)\,e^{-\alpha} + 2(N-2)\,e^{-2\alpha} + \ldots \\ + 2\,[N-(N-1)]\,e^{-(N-1)\alpha} .$$

Upon substituting $2N - N$ for N we obtain:

$$I \propto -N + 2N\,[1 + e^{-\alpha} + e^{-2\alpha} + \ldots + e^{-(N-1)\alpha}] \\ - 2\,[e^{-\alpha} + 2e^{-2\alpha} + \ldots + (N-1)\,e^{-(N-1)\alpha}] .$$

The first bracketed term is a finite geometrical series and equals

$$\frac{1 - e^{-N\alpha}}{1 - e^{-\alpha}} \equiv S .$$

Thus the last bracketed term is simply the derivative $-\,dS/d\alpha$, and we may write:

$$I \propto -N + 2\,(NS + dS/d\alpha) .$$

Now we make use of the fact that α is very small. If we designate $N\alpha$ as x, then

$$S = \frac{1 - e^{-x}}{x} N ,$$

References p. 271

since the denominator of the function S is approximately α. Then

$$\frac{dS}{d\alpha} = \frac{xe^{-x} - 1 + e^{-x}}{x^2} N^2 ,$$

and

$$I \propto - N + 2N^2 \left(\frac{-e^{-x}}{x}\right) - 2N^2 \left(\frac{xe^{-x} - 1 + e^{-x}}{x^2}\right) ;$$

upon dropping the term N, which is small compared with N^2, we obtain:

$$I \propto N^2 \frac{2}{x^2} \left[e^{-x} - (1 - x)\right] .$$

Expanded in Taylor series in x this becomes:

$$I \propto N^2 \left[1 + \frac{x}{3} + \ldots\right] .$$

With this equation, the theory provides a method of measuring average end-to-end distance of random coils, for the quantity x can be expressed in terms of R. Experimentally, one determines the slope at zero angle of I/N^2 as a function of s^2. This is equated to an expression in R which we obtain by substituting for x according to definitions previously introduced and then taking the indicated derivative. The definitions used are $x = N\alpha$, and $\alpha = 1/6(k^2s^2a^2)$; assuming that $R = \sqrt{N}a$, the expression for x is $1/6(k^2s^2R^2)$. Thus the equation is set up:

$$\left.\frac{d(I/N^2)}{ds^2}\right|_{s=0} \propto \frac{d}{ds^2}\left[1 - \frac{1}{3} \cdot \frac{1}{6} k^2s^2R^2 + \ldots\right] = - \frac{1}{18} k^2R^2 .$$

The experimental measurements required are intensity of scattered light at various angles (in order to determine slope at zero angle by extrapolation) and average molecular weight (to determine the number of links, N). Molecular weight is obtained from the same intensity data; needed constants are evaluated by measurements of a standard.

Two criticisms of this method should be mentioned. The first is that we have used a probability formula which is valid only for large N, and we have used it even for those calculations pertaining to particles separated by only a few links. The error involved is not great, however, because in the summation the value of α is so small compared to unity that only for large N does the value of $e^{-N\alpha}$ deviate significantly from unity. The second criticism is that the effect of excluded volume has not been considered. This constitutes a more serious error, but one which is inherent in the present state of the theory of random coils.

The usefulness of information such as average end-to-end distance depends on how it can be interpreted in terms of molecular structure and properties. As a specific example, let us suppose that we have calculated, from light scattering measurements, an average end-to-end distance of 1000 Å for a polystyrene sample of molecular weight 10^6, dissolved in benzene: what conclusions can we draw from this information? We know that the average molecule has a weight of $10^6 \times 1.64 \times 10^{-24}$ grams, and since its volume is of the order of R^3, its density is about 10^{-3} gm/cm^3. This result implies a very open structure, with the links of the random coil highly solvated, and indicates that benzene is a good solvent for polystyrene. In a solvent such as cyclohexane, R is found to be smaller, and this implies that cyclohexane is a poorer solvent than benzene, and that the dissolved molecules behave more independently.

Of course, as we have pointed out before, the application of such measurements is limited by the correctness of the underlying theory: the measurement is only as valid as the assumed molecular model. Conversely, the quality of a model is shown by how well and consistently it explains a variety of experiments. More detailed characterization of the structure of the macromolecule from light scattering data or, for that matter, from any physical data, will be possible with the development of a better general distribution theory of the random coil.

REFERENCES

1. EYRING, H., *Phys. Rev.*, 39, (1932) 746.
WALL, F. T., *J. Chem. Phys.*, 7, (1939) 289.
2. FLORY, P. J., *Principles of Polymer Chemistry*, Cornell Univ. Press, Ithaca, 1953.

INDEX